the safe and effective use of

FOG NOZZLES

Research and Practice

the safe and effective use of
FOG NOZZLES

Research and Practice

John D. Wiseman
John E. Bertrand

Copyright© 2003 by
PennWell Corporation
1421 South Sheridan
Tulsa, Oklahoma 74112 USA

800.752.9764
+1.918.831.9421
sales@pennwell.com
www.pennwellbooks.com
www.pennwell.com

Supervising Editor: Jared Wicklund
Cover and book design: Clark Bell
Cover photo: Robin Remaley

Cataloging-in-Publication Data Available on Request

ISBN: 0-87814-895-7

Printed in the United States of America

1 2 3 4 5   07 06 05 04 03

Dedication

This book is dedicated to Floyd W. (Bill) Nelson and Keith Royer. These two men, through their research at Iowa State University, were the first to learn how to use fog nozzles safely and effectively.

Table of Contents

List of Tables and Figures

Preface

This introduction is being written at the beginning of the 21st century, 50 years after almost all fire departments in the United States switched from using smooth-bore nozzles to fog nozzles. Yet arguments are being made today to return to smooth-bore nozzles. Andrew Fredericks, in his article "Little Drops of Water, 50 Years Later", concludes with the following statement: "Fifty years after Layman's 'Little Drops of Water', it's time to admit that fog nozzles are not the answer."[1] In a recent book published by the NFPA, authors Bernard J. Kleene and Russell J. Sanders conclude, "The reality is that there is little need for fog streams during offensive structural fire fighting."[2]

On the other side of the debate, Paul Grimwood notes how European fire departments have created and used a "new-wave" 3-D water fog attack. He says this is "...a new approach that has been evaluated, developed, and proven scientifically."[3] Thus, European fire departments have integrated the 3-D fog attack into their strategy and tactics so that fog nozzles play a crucial role in initial fire attack on almost all structure fires. Grimwood's argument is backed by Commander John P. Farley's (US Navy) report on a series of tests conducted at the US Naval Research Laboratory. These tests compared the effectiveness of the traditional straight-stream direct attack with the 3-D fog attack. Farley concluded the following: "... the offensive fog attack strategy is the best method to safely maintain a rapid, continuous, and aggressive response to a fire when entry to the fire can be made but direct access to the fire seat cannot be gained."[4]

The US Navy in 1994 officially approved and adopted the new 3-D fog attack for shipboard fire fighting.

This book explains the proper use of fog nozzles to provide the safest and most effective means of attack for certain types of fires. It also explains how fog nozzles have been misused, leading to the call by some for abandoning fog nozzles altogether. In our work we analyze the fire environment to determine the scientific facts and principles that govern fire behavior and water behavior. We also analyze more than 50 years of experience in using fog nozzles. However, neither experience nor trial-and-error methods alone necessarily lead to the correct conclusions.

Research is needed—research that involves very careful observations and systematic analysis, all of which is based upon a thorough knowledge of scientific facts and principles.

Thus, the heart of this book is the presentation of research projects dealing with the use of fog nozzles. Though most of the research was conducted in the United States, this work is not well known nor recognized. This lack of knowledge leads to the misuse of fog nozzles, and the calls to abandon fog nozzles completely. However, research tells us that fog nozzles can be used safely and effectively. Our book presents the strategy and tactics needed to use fog nozzles safely and effectively.

We acknowledge Paul Grimwood's help in writing the section on 3–D gas–cooling pulse tactics that are widely used in Europe. Grimwood, retired from the London Fire Brigade (England), is an author and authority on the use of fog nozzles, and he provided most of the information for the section on European research. A wealth of information is available on his Web site: www.firetactics.com. He can also be reached by e–mail at firetactics@aol.com.

We also acknowledge the cooperation of Stewart MacMillan, president of Task Force Tips. He freely made available the booklet published by Task Force Tips, "A Firefighter's Guide to Nozzles". Much of the information in chapter 8 is obtained from this booklet.

Notes

[1]Andrew Fredericks, "Little Drops of Water, 50 Years Later", *Fire Engineering* (Vol 153, No. 2, 3, Mar. 2000), p. 130.

[2]Bernard J. Kleene and Russell J. Sanders, *Structural Fire Fighting*, (Quincy, Mass., NFPA, 2000), p. 277.

[3]Paul Grimwood, "New Wave 3–D Water Fog Tactics", *Fire Engineering* (Vol 153 No. 10, Oct. 2000), p. 99.

[4]John P. Farley, "Fog Attack for Ship Fires", *Fire Engineering* (Vol 147 No. 3, Mar. 1996), p. 48.

CHAPTER 1

The Beginning

Fog nozzles have been in existence for more than 100 years. Several early fog nozzles were invented in the 19th century; by the 1930s, industrial fog nozzles were imported into the United States from Europe. These prim.tive nozzles changed the fog pattern by rotating the barrel, although this also changed the flow rate. As the United States produced more and more cars, trucks, and diesel locomotives, the refining, production, and use of petroleum products increased very rapidly along with the problem of fighting Class B fires. Solid–stream nozzles were worthless for fighting Class B fires; in fact, their use worsened the situation.

By 1925, the oil fields of Santa Fe Springs, California, had become well known as one of the worst fire problems in the United States. Battalion Chief Glenn G. Griswold was then captain of Engine Company 127, Santa Fe Springs Fire Station, in the Los Angeles County Fire Protection District. Because Captain Griswold had the advantage of being a hydraulic engineer, he began the task of creating new types of nozzles. Griswold created three types of fog nozzles, and by 1930 the California Fog Nozzle Company was producing these nozzles. These nozzles were later modified to become the U.S. Navy fog nozzles that are still in use today.

The Indirect Method of Attack

The first real research on the use of fog nozzles was conducted from 1943 to 1946 as a result of fires encountered during World War II. Chief Lloyd Layman, of the Parkersburg, WV, Fire Department became Commandant of the U.S. Coast Guard Fire Fighting School at Fort McHenry Training Station, Baltimore, Maryland. He conducted 20 experiments onboard a Liberty ship in its machinery, or engine, room. This space measured 50ft (15m) long, 53ft (16m) wide, and the height varied from 22ft to 32ft (6.7m to 9.7m). The net atmospheric volume (minus the machinery volume) was approximately 65,000ft³(1,840m³). Fuel oil, from 5,000gal to 7,000gal (18,925L to 26,495L) was released in the bilge space, providing approximately 1,800ft² (167m²) of burning surface. Most of the fuel oil was below the steel floor plates, thereby preventing any water from being applied directly to the fuel oil. The fuel oil was allowed to burn for at least 30min to provide the worst possible fire onboard this ship.

These experiments used the Navy's low–velocity fog nozzle, illustrated in Figure 1–1. The nozzle was attached to a short, 3ft (0.9m) applicator that was in turn attached to a 1.5in (38mm) hose line with 100psi (6.9bar) nozzle pressure. In the early tests, two lines were used that flowed about 168gpm (635Lpm). In later tests, only one line was used; it flowed 114gpm (431Lpm). The fog nozzles were lowered through a skylight near the aft bulkhead, allowing the water to be applied into the upper part of the machinery room some 28ft (8.5m) above the deck plates.

Closing the air–intake openings was the first and most important step taken in these experiments. This restriction in

Fig 1–1: Navy Low–Velocity Fog Nozzle with Applicator

oxygen supply reduced the fuel mass loss rate and minimized the rate of heat release. Only one opening could not be closed–the opening between the smokestack and the stack casing, with an area of 54ft² (5m²).

Immediately after the 30min burn when the fog attack began, there was an out rush of smoke from the exhaust opening. A mixture of smoke and condensing steam followed. Then this changed to condensing steam, and within a few minutes the amount of condensing steam began to decrease. Water continued to be applied for at least 25min to a maximum of 40min until no more condensing steam appeared. During this entire time the thermocouples recorded a continuous decrease in temperature throughout the machinery room to below 300°F (149°C). The fuel oil was cooled below its flash point of 200°F (93°C), with the temperature of the steel cooling to around 212°F (100°C).

Chief Layman realized that he had created a new method of attack that he called "the indirect method". He wondered how the burning fuel oil could be cooled and extinguished without applying water directly to the burning fuel surface. After considerable study, he reached the following two conclusions:

- Rapid generation of steam within the confined space creates a violent atmospheric disturbance within the space.

- Each cubic foot of steam generated within a confined space demands a cubic foot of that atmospheric space.[1]

Layman stated this "theory of indirect application and atmospheric displacement" as follows:

The cooling action of water in the form of finely divided particles at the upper atmospheric level within a highly heated confined space is not limited to the immediate area. The injection of water into a highly heated atmosphere results in rapid generation of steam, thereby creating an atmospheric disturbance of sufficient force to distribute unvaporized particles throughout the space. Unvaporized particles are brought into contact with heated materials located beyond the immediate area and at the same time contributing to the atmospheric disturbance by expanding into steam. It appears that this action continues until the surface temperature within the space is reduced to approximately 212°F (100°C), the boiling point of water.

Rapid generation of steam increases the atmospheric pressure within the space—each cubic foot of steam demanding $1ft^3$ of atmospheric space. A building is not air–tight, therefore, the interior and outside pressures are quickly equalized by an escape of atmosphere from the higher to the lower area. If the volume of steam generated within the space exceeds the net atmospheric volume of the space, most, if not all, of the original atmosphere will be displaced by steam. When the surface temperature within the space has been reduced to approximately 212°F (100°C), the boiling point of water, steam generation ceases.

At this time steam within the space starts to condense and cool air from the outside enters filling the void created by the process of condensation. This in draft of cool air from the outside atmosphere tends to increase the rate of condensation and continues until the process of condensation ceases. At this time, a major part, if not all, of the atmosphere within the space consists of normal air.[2]

The key to Chief Layman's explanation is the statement that unvaporized particles of water are blasted throughout the fire area and brought into contact with burning fuels throughout the space, thereby cooling the fuels and extinguishing the fire. If this were not the case, the indirect method would not work.

This new method of attack was an achievement of great historical significance since it was the first alternative to a direct attack using solid–stream nozzles. The term "indirect method" probably should be thought of in its simplest meaning, that is, as not the direct method of attack with solid streams then in use. Layman states that unvaporized particles are brought into contact with burning fuels located beyond the immediate area of application, thereby exerting cooling throughout the entire area. Thus, by steam expansion, water is brought into direct contact with all of the burning fuels. In reality, what Layman called an indirect attack is nothing more than a direct attack. Unvaporized particles of water are brought into direct contact with all of the burning surfaces by steam rather than by direct application from a fog nozzle. While it might seem to be contradictory to say that an indirect attack is really a direct attack, in essence that is what actually happens.

These experiments have been referenced repeatedly in subsequent years. The most frequent mention has been made of the fact that all but one of the air intake openings were closed. Many people have said that a fog attack (on land) would not work unless the structures were completely closed. However, one should be cautious in drawing any conclusions about structure fires based upon these early shipboard experiments because the circumstances were simply too different from what is found in structural firefighting. The volume (85,000ft³; 2,400m³) was much larger than the volume of an average–size house (20,000ft³; 566m³). The floor, walls, and ceiling of the machinery room were steel, capable of being sealed air–tight almost to the theoretical maximum. The low–velocity Navy fog nozzle had a reach of about 5ft (1.5m) in a semispherical shape in front of the nozzle. With a 22ft to 32ft (6.7m to 9.7m) ceiling height, no water was applied anywhere near the burning fuel oil surface. The net atmospheric volume of the machinery room required 325gal (1,230L) of water to completely fill that space full of steam. The water was applied at the rate of 114gpm to 168gpm (431Lpm to 635Lpm), requiring 2min to 3min to apply that amount. In fact, water was applied for an average of 30min, providing anywhere from 3,500gal to 5,000gal (13,247L to 18,925L) in these experiments. None of this data applied to structural firefighting, but it provided the knowledge on which further research was based.

As interesting as these experiments are, further experiments were needed to find out how to use fog nozzles safely and effectively for structural firefighting.

Parkersburg Experiments

Layman further concluded that much more work needed to be done to adapt his new method of attack to structural firefighting. When he returned to Parkersburg, West Virginia, as chief of the fire department, he experimented from 1947 to 1951.

Beginning in 1947, Layman took two years to train and equip the members of the Parkersburg Fire Department with the skills and equipment necessary to successfully use his new method of attack. Although Layman did not provide any details of this training in his book, the training certainly included the fundamentals of fire behavior and water behavior as presented in the first two chapters of his book.

Layman's most notable contribution here is his classification of three phases of structural fire development:

Phase I: Incipient or smoldering period
    Oxygen level = 21%

Phase II: Flame production period
    Oxygen level = 21% to 15%

Phase III: Smoldering period
    Oxygen level < 15%.

Chief Layman discussed in detail the equipment required to make a fog attack. The following nozzles were used.

- The Elkhart Mystery Nozzle—This nozzle was the industrial type with a rotating barrel that provided an adjustable fog pattern from straight stream to wide angle (60°). Even though the nozzle could be shut off by rotating the barrel, Layman used a Wooster Brass shut–off valve so the nozzle could be shut down quickly. At 100psi (6.9bar), one nozzle delivered 65gpm (246Lpm) fog and 60gpm (227Lpm) straight stream. The other nozzle delivered 90gpm (340Lpm) fog and 70gpm (265Lpm) straight stream.[3]

- The Elkhart Jumbo Mystery Nozzle—A master stream nozzle with a rotating barrel. Delivery rate 425gpm (1608Lpm) at 30° fog and 350gpm (1324Lpm) straight stream. (See Figure 1–3)[3]

- Rockwood Booster Nozzle–Flow rate on this nozzle was 18gpm (68Lpm) in a fixed fog pattern and 18gpm (68Lpm) as a ¼in (0.6cm) solid–stream nozzle. Impinging jets produced the fog. (See Figure 1–4)[4]

- Navy Low–Velocity Heads (Rockwood Sprinkler Co.)—These heads were attached to an applicator or a cellar pipe. One head delivered 114gpm (431Lpm) and the other head delivered 54gpm (204Lpm). (See Figure 1–1)

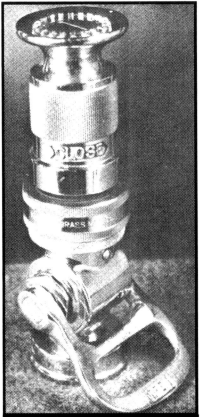

Fig 1–2: Elkhart Mystery Fog Nozzle

Fig 1–3: Elkhart Jumbo Mystery Nozzle

Fig 1–4: Rockwood Booster Nozzle

One interesting thing that Chief Layman did was to weld the Elkhart Mystery nozzles so the barrel could not be rotated to provide a fog pattern wider than 30°. He believed this setting provided the optimum reach and the best fog, that is, the best size of water droplets. All of his firemen were instructed to use the 30° fog pattern when making a fog attack.

However, Chief Layman recognized that the equipment problem had not been solved with the use of these nozzles. So he added the following:

> This system of firefighting offers a real challenge to inventors, designers, and manufacturers of firefighting equipment. They should approach this problem with a comprehensive understanding of the natural laws that govern the extinguishing action of water and its proper tactical employment on the fire ground. Progress in the art of firefighting demands practical equipment that will enable the fire service to utilize the vast extinguishing action of water more effectively in attacking and extinguishing fire.[5]

Thus Chief Layman's pioneering research established a foundation for safe and effective use of fog nozzles. He looked forward to further research that would establish the most effective use of fog nozzles.

Case Histories

In his book, Layman presented six case histories. Detailed reports and photographs were made on all working structure fires in Parkersburg. To give a clear picture of how the fog attack was conducted, the first case history is presented here in some detail. The other five case histories are also summarized.

Case History 1

This fire occurred in a one–story dwelling in Parkersburg in May 1949. The 1st floor of 1,800ft^2 (167m^2) was divided into two apartments from front to rear with a short common hallway and a common bath at the rear. There was a semi–finished basement with an interior stairway and another inside stairway leading to the attic from the west–side apartment where the fire occurred. The west–side apartment also had a small rear porch with no access to the ground. The fire started in the kitchen from an over-heated refrigerator electric motor.[6] (Insert Figure 1–5)

Fig 1–5: 2000 16th Street

Upon the arrival of the first fire truck, the kitchen and rear porch were involved in fire. Heavy smoke filled the west–side apartment and the attic. Firefighters broke the west–side window and made their initial attack. Using a ladder to gain access to the porch, they made a direct attack on the rear porch. Chief Layman described the initial attack as follows: "Immediately there was a violent expulsion of smoke followed by a mixture of smoke and condensing steam. The entire house was enveloped in a cloud of smoke and condensing steam."[4] Two 1.5in (38mm) lines were used in the attack, both with high–velocity fog nozzles flowing approximately 54gpm and 65gpm (204Lpm and 246Lpm). Less than 200gal (757L) of water were used to extinguish the fire. Approximately

125gal (473L) were used in the attack through the kitchen window, which lasted about 2min. Approximately 75gal (284L) were used in the attack on the exterior fire on the porch, leaving less than 5gal (19L) of water on the kitchen floor and a small area of the bedroom floor.

The volume of the west–side apartment and the attic was 11,000ft^3 (311m^3). The 125gal (473L) of water applied through the kitchen window produced approximately 25,000ft^3 (707m^3) of steam based upon a 90% conversion rate. This was enough steam to replace the inside atmosphere of the apartment two times. Needless to say, the conclusion drawn by Chief Layman is fully justified: "The most practical and effective method of controlling and extinguishing this type of fire is by the proper application of the necessary volume of water in the form of finely divided particles."[5]

This 1949 fire was the first recorded use of water fog in extinguishing a Class A fire. As with Class B fires, the results were spectacular. The fire was extinguished with a very small amount of water in a very short period of time.

A conclusion from this case history is that a fire does not have to be completely confined for a fog attack to be effective. This fire had burned through to the outside rear porch and had also burned through the rear attic window.

Case histories 2–6

The second fire occurred in a vacant two–story house with an unfinished attic with several gables. The initial attack was made from the inside stairway at the top of the stairs. Fifty gallons (189L) of water were used, being converted to 10,000ft^3 (283m^3) of steam. The volume of the attic space was 4,000ft^3 (113m^3). Again, this volume of steam was enough to have two complete changes of atmosphere in the attic. The nozzle operator was forced to retreat from his position at the top of the stairs, although he returned and used an additional 50gal (189L) of water to control the attic fire. The third fire occurred in a warehouse that was divided into two sections—a west section of 30,000ft^3 (849m^3) net volume and an east section of 27,000ft^3 (764m^3). The ceiling height was 16ft (4.9m) on the west end and 19ft (5.8m) on the east end. The fire started from a rubbish pile at the rear of the building that spread into the warehouse and burned through the roof at the rear. Four 1.5in (38mm) attack lines were used to extinguish the fire. One line was used exclusively to protect exposures and to extinguish the trash fire at the rear. Two lines were used to make a direct attack through two of the four burned–out windows in the east section. The fourth line was used to make an indirect attack through the transom of a door leading into the west section. The fourth line used 270gal (1,021L) in the indirect attack that converted to 54,000ft^3 (1,528m^3) of steam. The volume of the west section was 30,000ft^3 (849m^3) net. Again, this is very close to two complete changes of atmosphere for this section of the warehouse. Using four lines, the attack successfully extinguished the fire.

The fourth fire was in an ordinary one–story dwelling in which fire involved three rooms with heat spreading to the other rooms. The net volume of the house was 8,400ft^3 (238m^3). The initial attack was made at the rear bedroom window and then progressed to the rear porch where an attack was made through the kitchen door. About 85gal (321L) were used, which produced 17,000ft^3 (481m^3) of steam. That volume was sufficient to provide two complete atmospheric changes within the house. The fire was extinguished using a single attack line.

The fifth fire occurred in the basement of a small house. The basement measured 33ft × 24ft (10m × 7m) with a net atmospheric volume of 3,500ft^3 (99m^3). There were not very many combustibles in the basement. The fire started when a 1gal (3.8L) jug of gasoline was dropped, resulting in gasoline vapors being ignited by a pilot light. The initial attack was made through a basement window with a 1.5in (38mm) attack line flowing 65gpm (246Lpm) for 20s, thereby flowing 20gal (75L) of water. Twenty gallons vaporizes to 4,000ft^3 (113m^3) of steam. This volume just about equals the volume of the basement. This fire was extinguished very quickly with little damage to the house.

Finally, the sixth fire occurred in a commercial building with a three–story section in front and a one–story section behind. The fire started on the 1st floor near the elevator shaft and spread upward through the 2nd and 3rd floors and out through the roof. The attack was made with two 2.5in (64mm) lines with solid–stream nozzles flowing 220gpm (832Lpm) each. The first line quickly knocked down the fire on the 1st floor at the base of the elevator shaft. The second line was taken to the roof of the

one–story section and was used to attack the fire through a second–story window. The window had heavy metal bars that could
not be removed. Thus, the solid stream was broken up to some
extent by these bars.

The attack through the 2nd floor window blacked out the
fire, and the nozzle was shut down. Then the fire reappeared and
the nozzle was opened again; the flames were extinguished. This
attack continued for about 7min, flowing about 1,500gal (5,678L)
of water. The net volume of the 2nd floor was 25,000ft³ (707m³).
The net volume of the 3rd floor was 33,000ft³ (934m³). Vaporization of 125gal (473L) was needed for the 2nd floor, and 165gal
(624L) were needed for the 3rd floor. Layman estimated that anywhere from 300gal to 500gal (1135L to 1892L) were vaporized out
of the 1,500gal (5,678L) applied.

Chief Layman was dissatisfied with this attack—particularly with the amount of water used. He concluded these things:

1. A single 1.5in (38mm) line equipped with a 65gpm (246Lpm)
   high–velocity fog nozzle would have provided sufficient
   rate of application to have extinguished the fire on the
   first floor.

2. A single 1.5in (38mm) line equipped with a 90gpm (340Lpm)
   high–velocity fog nozzle would have provided a sufficient rate of application to have displaced the original
   atmosphere and extinguished surface burning on 2nd
   and 3rd floors and in the cockloft, provided the cone of
   water particles had been directed into the upper stratum
   of the interior atmosphere through the burned–out
   window.

3. If an indirect attack had been made through the burned–
   out window using a 90gpm (340Lpm) high–velocity fog
   cone, equal or better results would have been obtained
   with one–third to one–fifth the volume of water that was
   applied with the solid streams.[7]

Three conclusions can be drawn from these six case histories. First, there is a definite relationship between the volume of
water applied and the net atmospheric volume: 2:1 in four of

these fires and 1:1 in the fifth Class B fire. In the sixth fire, much more water was applied through solid–stream nozzles than was needed to fill the net atmospheric volume full of steam.

Therefore, an indirect attack probably requires a volume of water that provides two atmospheric changes in the fire volume, but only one atmospheric change is needed for fire control. If the original contaminated atmosphere is displaced by steam, then as the steam condenses the inflow of cooler air means the atmosphere consists of normal air. So a second displacement is not needed. In fact, there is no way for a fog nozzle to displace normal air with steam because normal air is not hot enough to produce steam. What is needed is a fog attack that uses a volume of water equal to the volume of the fire.

Second, in five of the six case histories only one line was used to control the structure fire. This is true of the commercial building since the fire line was used only briefly on the 1st floor. The only exception is the warehouse fire. In four cases, fire was extinguished in adjacent rooms, or spaces, without any water being applied to these areas. This is called the indirect effect of the indirect attack. The indirect effect is especially effective in areas above the fire floor, which explains why the indirect effect is not limited to the immediate area, or room, where the water is applied. Remember Layman's explanation of why this happens. Unvaporized particles of water are blasted throughout the fire area, including adjacent areas, so that little drops of water are brought into contact with burning fuels throughout the space.

Finally, the flow rate needed to control a one–room structure fire is well below 100gpm (378Lpm). Layman's data indicate that flow around 50gpm (189Lpm) is sufficient for such fires. One 1.5in (38mm) fully open attack line provides too much water for such a fog attack.

Chief Layman summarized the fundamentals of the indirect method of attack as follows:

1. Degree of confinement and concentration of excessive heat are important factors in this method of attack.

2. The initial attack should be made within the area of major involvement.

3. Atmospheric displacement will occur only on and above the floor where an indirect attack is made.

4. The progress of an indirect attack can be estimated readily by observing the volume of smoke and condensing steam coming from the building.

5. Injection of water particles should continue without interruption until the volume of condensing steam coming from the building has decreased to a major degree.

6. After an indirect attack has been executed in the proper manner, personnel may enter and operate within the building—perhaps with some discomfort due to high humidity rather than high heat.

7. An indirect attack should always be made from a position that will enable personnel to avoid injuries from heated smoke and steam.

8. When openings are made, they should be small; a doorway is an undesirable type of opening.

9. If the attack is made through a window or other low–level opening or from an interior stairway, a high–velocity cone should be used.

10. A fire in a cockloft or unfinished attic may be attacked by making a small opening in the ceiling below and inserting a straight applicator fitted with a low–velocity nozzle.

11. Where an attack must be made from a floor above, a cellar pipe should be used; personnel should be withdrawn before the attack is made.

12. A high degree of dispersion of water particles within the upper stratum of the interior atmosphere is an important factor in obtaining rapid heat absorption and a high percentage of conversion.

Layman added to point 12 that considerable skill and confidence are required to obtain a high degree of dispersion when using high–velocity nozzles in an indirect attack. Besides recognizing that the equipment problem had not been solved, Layman also recognized that the most effective flow rate posed a difficult issue for which there is no definite formula. He was limited, of course, by having nozzles whose rate of flow varied with the rotation of the barrel.

Many statements have since appeared in print that do not accurately reflect Layman's views. He did not advocate the exclusive use of the indirect method of attack. He said that fires in the first phase of development have not generated enough heat to justify an indirect attack. He clearly stated that such fires must be located and extinguished by a direct attack. He added, "An experienced and capable officer should have little difficulty in determining if the situation demands a direct or indirect attack."[7] Chief Layman discouraged the use of booster lines in making an indirect attack. His reason: "The volume of water per minute is too small to produce satisfactory results in larger spaces except under the most favorable conditions."[8]

Chief Layman also said that he was frequently asked a question about the indirect method of attack: "What is the effect of steam from an indirect attack upon occupants in a building?" Chief Layman said that he had never heard of any adverse effects. He added, "The much more rapid flame suppression with indirect application makes it possible to reach endangered persons more quickly so as to be able to remove them to safety and render aid as necessary."[9]

1950 "Little Drops of Water"

By 1950, Layman's work had come to the attention of a number of officials in the fire service. Officials of the National Fire Protection Association (NFPA) became interested, and they later published two books by him. Also, officials of the Western Actuarial Bureau, Chicago, Illinois, decided to explore the use of fog nozzles. As a result, Layman was invited to address the annual Fire Department Instructor's Conference in Memphis, Tennessee. His speech marked an historic turning point for the fire service in the United States. On January 11, 1950, he gave a speech entitled "Little Drops of Water", promoting his new method of fog attack. He concluded that "Little if any progress can be made in improving the tactical employment of water in firefighting operations until the gross ineffectiveness of the solid stream method of application is recognized."[11]

Chief Layman's speech was not only heard by the attendees at the FDIC Conference, but also it was read the following month in Firemen magazine. The NFPA reprinted this article and widely circulated it.

The Exploratory Committee on the Application of Water

The Advisory Engineering Council of the National Board of Fire Underwriters met in Detroit in May 1951. R.D. Hobbs, manager of the Western Actuarial Bureau (W.A.B.), authorized Richard E. Verner, manager of the W.A.B. fire prevention department, to appoint a committee to conduct tests of the new indirect method of attack. In 1951, Verner appointed a committee known as the Exploratory Committee on the Application of Water. Emmett T. Cox of the W.A.B. was named chairman, and the committee was comprised of 33 members who represented fire insurance, engineering, inspection, rating organizations, the fire service, and fire training agencies.

The Exploratory Committee selected Miami, Florida, as the site where the first tests would be conducted. Chief Henry Chase of the Miami Fire Department was a member of this committee; his department had been using fog nozzles for some time. Also, the Miami Fire Department had a fully instrumented building available. The tests were conducted from October 29 through November 2, 1951. Interestingly, the Miami Fire Department provided a real fire as well for the members of the committee to investigate. On October 28, a fire occurred in an automobile repair garage, completely involving the $4,800\text{ft}^2$ (446m^2) structure (one story) before the alarm was sounded. Two 1.5in (38mm) lines with adjustable fog nozzles controlled this fire. The members of the committee found this very instructive since the fire was controlled quite effectively by the use of fog nozzles.

The test building was two stories of fire–resistive construction featuring concrete walls, ceiling, roof, and windows with movable concrete shutters. The building had 17 fixed thermocouples—eight on the 1st floor and nine on the 2nd floor—with an 18th movable thermocouple. Wooden partitions divided the 1st floor into eight rooms, and each room had a thermocouple. The fog nozzles used had different gpm capacities but were usually set at a 30° fog pattern. Nozzle pressure was 100psi (6.9bar), and the fuel load was approximately 10lb/ft² (4.5kg/m²).

One purpose of the tests was to verify the indirect extinguishing effect of fog nozzles. A second purpose was to identify training problems in teaching the new method. The tests were not designed to determine why the indirect effect was produced or to develop techniques for maximum effectiveness.

Four tests were held, with the first two tests involving one room on the 1st floor. Test 3 involved the entire 1st floor, and test 4 involved the entire 2nd floor. Test 4 utilized a greater flow rate that produced much better results than test 3. Only one attack line was used on each test. Table 1–1 summarizes the results of these tests.

Table 1–1: The Miami Tests

Test	Time (min)	Flow Rate		Water used	
		(gpm)	(Lpm)	(gal)	(L)
1	11.0	28	106	308	1,165
2	4.8	85	321	408	1,544
3	19.9	85	321	1,692	6,404
4	7.5	180	681	1,350	5,109

The committee did not draw any conclusions about the time, flow rate, or amount of water used. Their conclusions were related to the original purpose of the tests:

1. In all of the tests, the indirect effect of the fog or spray streams was clearly apparent. The water vapor penetrated all parts of the structure.

2. In one demonstration, the Miami Fire Department extinguished a fire using new men who had just completed basic training. Based upon this, the committee maintained it was unnecessary for firefighting forces to have extensive training to use the technique properly.

3. The results of the tests and demonstrations indicated sufficient effectiveness of water fog or spray to warrant additional study and tests under other circumstances.

4. Since extensive training in the use of the technique to prevent increased fire was proven unnecessary, the committee concluded fire departments should be encouraged to use the method when opportunity was presented.

The report ended with a broad range of recommendations for further study. Still, given the limited objectives of these Miami tests, further testing was needed to establish the most effective methods for using fog nozzles.

In an attempt to make the tests more realistic, in 1952 a subcommittee of the Exploratory Committee directed a significant series of tests in Kansas City, Missouri. This was the first time that tests were conducted in dwellings with common furnishings. The three test buildings were small: one 23ft × 32ft (7m × 10m) and one 22ft × 30ft (7m × 9m) one–story structures and a two–story 22ft × 40ft (7m × 12m) structure. The construction was frame and stucco, or block walls, with lath and plaster interior walls and ceilings.

All of the tests used a 1in booster line with a Rockwood fog nozzle. In three of the tests, two booster lines were used. The initial attack was followed up by a direct attack to extinguish the fire. These tests represented real progress in establishing the effectiveness of fog nozzles with a time of 60s or less for a one–room fire, a flow rate less than 50gpm (189Lpm), and less than 50gal (189L) of water used.[11] Table 1–2 shows the test results.

Table 1–2: The Kansas City Tests

Test	Lines	Time (s)	Flow Rate (gpm)	Flow Rate (Lpm)	Water used (gal)	Water used (L)
1	1	20	25	94	8.0	31
2	1	180	25	94	75.0	282
Attic	1	42	25	94	17.5	66
3	2	80	50	189	66.0	250
2nd Floor	2	60	50	189	50.0	189
1st Floor	2	180	50	189	150.0	567

The Exploratory Committee conducted or supervised about 40 more tests or demonstrations during the next seven years. In addition, individual members of the committee attended and reported on more than 100 other tests. State fire schools or individual fire departments conducted many of these tests. The result of all of this activity was the widespread adoption of the indirect method of attack and the almost universal use of fog nozzles.

As fog nozzle acceptance increased, members of the Exploratory Committee also participated in a series of nozzle tests from 1952 through 1959 sponsored by the NFPA and the International Association of Fire Chiefs (IAFC). A photographic subcommittee of the Exploratory Committee also produced four films: "Using Water Wisely", "Master Spray Stream Problems", "Fog Against Fire", and "Let's Try Fog". Members of the photographic committee also assisted in producing films by other organizations.

The Exploratory Committee certainly did its job thoroughly. Not only did it explore the new indirect method of attack and the use of fog nozzles, but it also exerted a profound influence upon numerous organizations throughout the United States.

Summary

Substantial progress was made from 1943 to 1959 in establishing the effectiveness of fog nozzles. In 1943, experiments were conducted onboard a single Liberty ship at the Coast Guard Training Station in Baltimore, Maryland. By 1959, tests and demonstrations had been held throughout the United States by the Exploratory Committee. The result of this activity was that almost all fire departments switched from using solid–stream nozzles to using fog nozzles in fighting structure fires. This was truly a revolutionary change for the fire service.

The first experiments at the Coast Guard Fire School stand in stark contrast to the subsequent experiments. The 20 experiments onboard the Liberty ship had a typical application time of 30min, a flow rate of 114gpm or 168gpm (431Lpm or 636Lpm), and used 3,500gal to 5,000gal (13,247L to 18,925L) of water. If these numbers had truly measured the effectiveness of fog nozzles, then that would have been the end, not the beginning.

The Parkersburg and Kansas City experiments advanced the use of fog nozzles. A time of 1min to 2min was typical, depending upon whether one or more rooms were involved in a fire. The flow rate varied from 65gpm to < 100gpm (246Lpm to < 378Lpm) with 50gal to 100gal (189L to 378L) used. The Kansas City tests used a booster line that had a flow rate of 25gpm (94Lpm). These tests indicated that for a one–room fire, a fog nozzle could be used to control a fire in seconds with a flow rate of 50gpm (189Lpm) and using less than 50gal (189L) of water. Still, these tests do not establish an exact mathematical relation between the volume of the fire and the volume of water used.

The Parkersburg tests clearly indicate that the indirect method of attack was not as effective as it should have been. The fact that a volume of water was used that equals two net atmospheric changes indicates an effectiveness of 50% for the indirect method. However, these tests do establish an indirect effect for the use of fog nozzles. That is, the rapid steam expansion affects not only the room where water is applied but also adjacent areas that have been heated more than 212°F (100°C). This atmospheric displacement increases the effectiveness of the attack and greatly shortens the time of attack since one line from one position can control a fire in adjacent areas. Thus, it is not necessary to move the nozzle to different positions or to use several attack lines at different positions.

The key is Layman's explanation that unvaporized drops of water are blasted throughout the fire area, including adjacent areas, thereby cooling the burning fuels. This new method of attack provides an indirect application of water in one location and then steam blasts unvaporized particles of water throughout the entire fire area. The water, in turn, directly cools the remaining fuels. In essence, an indirect attack is a direct attack throughout the fire area. If you understand why this statement is true, then you understand Layman's theory of the indirect attack and atmospheric displacement.

This historic research provides us with the foundation for more modern inquiry, giving us the basic format for new research. Modern researchers, as we will see, have produced excellent findings using Layman's experiments as a starting point.

Notes

[1]Lloyd Layman, *Attacking and Extinguishing Interior Fires* (Quincy, Mass., NFPA, 1955), p. 36.

[2]Ibid, p. 36 ff.

[3]Ibid, p. 58.

[4]Ibid, p. 59.

[5]Ibid, p. 70.

[6]Ibid, p. 70.

[7]Ibid, p. 138.

[8]Ibid, p. 40.

[9]Ibid, p. 145.

[10]Ibid, p. 146.

[11]John D. Wiseman, *The Iowa State Story,* (Stillwater, OK, Fire Protection Publications, 1998), p. 46

This chapter describes the successful research at Iowa State University that began in 1951. During the same time many fire departments that had switched to fog nozzles were encountering serious problems. First, let's examine the research done at ISU.

In 1951, Iowa State University hired Keith Royer and Floyd W. (Bill) Nelson to manage the Firemanship Training Program that was part of the Engineering Extension Service. Both men became involved with the Exploratory Committee on the Application of Water, and Royer attended the first tests conducted at Miami. They quickly agreed on the need for independent research beyond that being done by the Committee. Many questions were being asked about the use of fog nozzles, but there were very few answers.

Fire Behavior

The first question they asked was what happens in a structure when a fire occurs. At this time there was little knowledge about fire behavior. Most of the information available was from combustion engineering, and it was not very helpful in understanding uncontrolled structure fires. After years of work, the two men arrived at the following analysis of fire behavior, broken into seven stages:[1]

1A.   Ignition—Fire becomes self-sustaining

1B.   Early Explosion—Usually ends further fire development

2A.   Flame Spread—Rapid buildup of fire intensity and volume

2B.   Cool Smoldering—Slow buildup of heat, temperature < 1,000°F (537.7°C)

3.   Hot Smoldering—Oxygen level below 15%, temperature ranges from 1,200°F (648.8°C) to more than 1,800°F (982.2°C)

4.   Flashover—Brief but spectacular state in which the area becomes fully involved; occurs at 1,000°F (537.7°C) ceiling temperature

5.   Steady State—A partially or fully open fire with ample fuel and oxygen, increasing severity, and temperatures between 1,400°F and 1,450°F (760°C and 787°C)

6.   Clear Burning—Smoke clears, peak severity, temperature > 1,500°F (815.5°C)

7A.   Post-Attack Warm—warm [300°F (148.8°C)], thermal balance, easy overhaul

7B.   Post-Attack Cool—Lower temperature, thermal imbalance, difficult overhaul

With specific reference to stages 7A and 7B, Royer and Nelson soon discovered that how a fog nozzle is used makes a difference. If it is effective, thermal balance is preserved and overhaul is easy. Otherwise, thermal imbalance is created and overhaul is difficult.

In his book *Qualitative Fire Behavior*, Bill Nelson presents a time–temperature graph for these seven fire stages, shown here as Figure 2-1.[2] The stages are placed on the graph at their approximate temperature level, using the average ceiling temperature as a guide. The time scale is indefinite because a fire may progress rather rapidly through the flame spread stage or rather slowly through the cool smoldering stage. Also, although fire development is usually from left to right on the graph, from flame spread to flashover, at times a fire may move from right to left, from flashover back to cool smoldering. This back-and-forth movement is called *oscillation*.

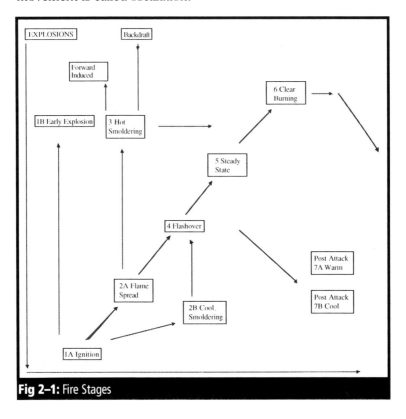

Fig 2–1: Fire Stages

Every firefighter should be familiar with these seven fire stages. Royer has emphasized the importance of every firefighter being a good student of fire behavior. There is no better place to start than Bill Nelson's book, *Qualitative Fire Behavior.*

The Fundamental Principle

Royer and Nelson confronted a second basic question: How much water is needed to control a fire with fog nozzles? The answer to this deceptively simple question eluded Layman for years. The answer is critical since it is directly related to the fundamental principle stated in Nelson's book, *Qualitative Fire Behavior:*

> In principle, firefighting is very simple. All one needs to do is put the right amount of water in the right place and the fire is controlled.[3]

This is one of the most profound statements ever made about fighting fires. Notice the phrase, "all one needs to do". Nothing else is needed, except using the right amount of water.

Why is using the right amount of water so important? The answer lies at the heart of what happens when water is used to fight fires. Fire is basically an exothermic, heat-releasing chemical process. Layman saw heat as the main problem that confronts firefighters. He noted, "The control and extinguishment of interior fires must be based upon the principle of removing excessive heat from the involved building".[4] So how do we fight fire with water? We do this by the endothermic, heat-absorbing process of changing liquid water to steam. In essence, the fundamental principle says that we must balance the exothermic process with the endothermic process of vaporizing water by using the right amount of water. It is true, as Royer and Nelson discovered in their research, that having too little or too much water is equally less effective. The maximum effective use of fog nozzles requires that the right amount of water be used.

How is the right amount of water calculated? Royer and Nelson answered this question by creating the following formula:

$$gal = \frac{Vol}{200}$$

In this equation

gal = the number of gallons (the right amount of water)

Vol = the volume of the confined space in cubic feet

200 = the constant determined by two scientific facts

Royer explained as follows how this formula emerged from their research.

1. Study of expansion ratios of water to steam indicates that 1gal of water will produce with a margin of safety $200ft^3$ of steam.

2. Study of heat production in relation to oxygen also indicates that in the conversion of water to steam, 1gal of water will absorb with a margin of safety all of the heat that can be produced with the oxygen available in $200ft^3$ of normal air.

These two factors lead to the formula that the cubic area of a room (in feet) divided by 200 equals the required number of gallons of water to control a specific area involved in a fire.[4]

The proof of the validity of this gallonage formula depends upon the constant 200. The proof follows.

Fact One is that the expansion ratio of liquid water to steam at 212°F (100°C) is 1:1,700. Since $1ft^3$ of water contains 7.48gal,

$$\frac{1,700}{7.48} = 227$$

That is, 1gal of water produces $227ft^3$ of steam. This number is rounded down to 200 to allow for 90% efficiency in conversion to steam.

Fact Two is that in 1955 the Factory Mutual Laboratories determined that $1ft^3$ of pure oxygen combined with ordinary fuels produced 535Btu of heat. Air contains 21% of oxygen by volume, and flame production stops when the oxygen level falls below 15%. Therefore,

$$21\% - 14\% = 7\%$$

Only this amount of oxygen, 7% of air, is available for flaming combustion. Multiplying this number by the number of British thermal units produced by 1ft^3 of pure oxygen gives

$$535 \times 0.07 = 37\text{Btu}$$

This is the number of British thermal units produced by 1ft^3 of air. Therefore, the British thermal units produced by 200ft^3 of air is

$$37 \times 200 = 7{,}400\text{Btu}$$

To raise the temperature of 1gal of water from 62°F to 212°F requires 1,250Btu. To vaporize 1gal of water at 212°F requires 8,080Btu. If 1gal of water is applied to a fire at 62°F, then the gallon absorbs 9,330Btu. Since 7,400 is less than 9,330, the conclusion is that 1gal of water absorbs all of the heat produced by 200ft^3 of air. Notice the margin of safety—almost 2,000Btu.

It is quite remarkable that both scientific facts produce the same constant, 200. This provides a solid foundation for the validity of the gallonage formula.

Metric Section

When moving from English to metric units, the gallonage formula changes to the liter formula. The liter, like the gallon, is a measure of volume, with 1gal = 3.785L. However, the volume of structures is usually expressed in cubic meters. One liter equals one cubic decimeter, or one one-thousandth of a cubic meter. In other words, 1,000L equals = 1m^3.

Even though we are changing units of measure, the expansion ratio of liquid water to steam remains constant, that is, 1700:1. One liter expands to 1,700L of steam. To change to cubic meters, it is necessary to divide 1,700 by 1,000. This quotient, 1.7, is the new constant for the liter formula for the right amount of water.

$$L = \frac{\text{Vol}}{1.7}$$

The Iowa Rate-of-Flow formula has a margin of error of 10% by rounding 227 down to 200. This assumes that 90% of the liquid water is transformed into steam. To make the same change for the metric formula we must use 1.5, that is, 90% of 1.7:

$$\text{NFF} \times t = \frac{\text{Vol}}{1.5}$$

In this equation

NFF = needed fire flow, measured in Lpm (gpm in American units)
t = time, measured in minutes

In this formula, volume is measured in cubic meters. The Iowa formula in liters then becomes

$$\text{NFF(30}_\text{s}) = \frac{\text{Vol}}{0.75}$$

The rate formula, in liters, is

$$\text{NFF} \times t = L$$

The proof of the validity of General Rate-of-Flow formula depends upon two scientific facts. The first is the expansion ratio of liquid water to steam at 212°F (100°C). This ratio is used to derive the constant, 1.7. The change involves the definition of a liter (L), so volume is expressed in cubic meters.

The second fact is that 1gal of water absorbs all of the heat produced by the oxygen in 200ft³ of air. One cubic foot = 28.3L, so 200ft³ = 5,660L. We know that 200ft³ of air produces 7,400Btu. We also know that 7,400Btu = 7,807,000J (joules). Therefore,

$$5,660\text{L} = 7,807,000\text{J}$$

The transformation of 1gal of water to steam requires 9,330Btu. If 1gal = 3.785L, then 3.785L absorbs 9,843,150J. Since 7,807,000 < 9,843,150, this gives the same proof as calculated in the English system of measure.

The Iowa Rate-of-Flow Formula

Based upon their research, Keith Royer and Bill Nelson concluded that almost all fires could be controlled in less than 30s by using fog nozzles. This fact was used to create what has become known as the Iowa Rate-of-Flow formula. Let's start with the gallonage formula:

$$\text{gal} = \frac{\text{Vol}}{200} \qquad \text{gal} = \frac{\text{Vol}}{1.5} \text{ (metric)}$$

The right amount of water can be applied to a fire at different rates of flow for different lengths of time. So let's add the rate formula:

$$\text{NFF} \times t = \text{gal} \qquad \text{NFF} \times t = \text{L (metric)}$$

Since both of these equations equal the same number (gallons), they can be combined by substitution to produce the General Rate-of-Flow formula:

$$\text{NFF} \times t = \frac{\text{Vol}}{200} \qquad \text{NFF} \times t = \frac{\text{Vol}}{1.5} \text{ (metric)}$$

Since the Iowa Rate-of-flow formula is valid for 30s, we want to substitute t = 0.5 (30s = 0.5min). It is important to note that this is the only tricky part of the formula. Time must be expressed in minutes, or a fraction of a minute, since NFF is in gallons per minute. If 30s were used instead, the results would be nonsense.

So substituting $t = 0.5$ into the General Rate-of-Flow formula gives the following equation:

$$\text{NFF} \times 0.5 = \frac{\text{Vol}}{200} \qquad \text{NFF} \times 0.5 = \frac{\text{Vol}}{1.5} \text{ (metric)}$$

Dividing both sides of this equation by 0.5 gives

$$\text{NFF} \times \frac{0.5}{0.5} = \frac{\text{Vol}}{1.5} \times 0.5 \qquad \text{NFF} \times \frac{0.5}{0.5} = \frac{\text{Vol}}{1.5} \times 0.5 \text{ (metric)}$$

Simplifying the equation by dividing and multiplying produces the Iowa Rate-of-Flow formula.

$$NFF = \frac{Vol}{100} \qquad NFF = \frac{Vol}{.75} \text{ (metric)}$$

Notice that t has disappeared from this equation. This is highly unfortunate since the Iowa formula is valid for only 30s (0.5min). The Iowa Rate-of-Flow formula should never be used for any purpose without taking into consideration the time of 30s.

You could tack the phrase "for thirty seconds" after the equation, but a better solution is to incorporate time into the equation. Function notation enables you to do this easily. Function notation looks like this:

$$f(x)$$

The expression $f(x)$ is read "f of x". This indicates that the value of f depends upon the value of x and, hence, is a function of x. The rate formula in function notation becomes

$$NFF\ (t)$$

This indicates that the value of NFF depends upon the value of t. Since the Iowa formula depends upon a time of 30s, this formula expressed in function notation becomes

$$NFF(30_s) = \frac{Vol}{100} \qquad NFF\ (30_s) = \frac{Vol}{.75} \text{ (metric)}$$

Using this notation would have corrected serious problems that occurred with the almost universal misuse of the Iowa formula. The authors have never seen the Iowa Rate-of-Flow formula used in which it was recognized that the formula is valid only for 30s. For an explanation of this problem, see the section on "The Identity Error" later on in this chapter.

One more error has hampered the correct use of the Iowa formula. The formula should be used only for the largest open area of a building, not for the entire building—unless, of course, the building itself is one large open area. Thus, the Iowa formula cannot be used with or compared to other formulas that are based upon the volume of a building or its square foot area.

Royer has emphasized repeatedly that the most important use of the Iowa formula is for preplanning. Choosing 30s as the standard for judging a fire department's ability to control a structure fire is certainly valid. No problems arise from this use of the Iowa formula.

However, it is a great mistake to use the Iowa formula for fire attack. The formula was not designed for this purpose. The reason why can be explained by the following example. An average room is about 2,000ft^3 (56.7m^3). By the Iowa formula,

$$\text{NFF (30s)} = \frac{2,000}{100}$$

$$\text{NFF (30s)} = 20\text{gpm}$$

This immediately creates a problem. Are you going to flow 20gpm (75.7Lpm) for 30s to fight a room-size fire? We don't think so.

First, such a fire can be controlled (knocked down) much faster than 30s. If you can do it quicker, you should do so.

Second, how could you flow 20gpm using the attack lines and fog nozzles that we have today? How about using a booster line (0.75in or 19mm) that flows 25gpm (94.6Lpm)? We don't think anyone would recommend going back to using booster lines for initial fire attack. (We certainly do not.)

What is needed is the General Rate-of-Flow formula that allows you to apply the right amount of water while varying the rate of flow with different lengths of time. The statement is not meant to diminish the importance, or significance, of the Iowa Rate-of-Flow formula. Its creation is certainly an important historical event for the fire service. As Keith Royer has said:

> The Rate-of-Flow formula is the only valid tool that we have that allows you to analyze a structure, before the fact, how much water it is going to take to control that— whether it is manageable or unmanageable from a fire control point-of-view.[6]

Thermal Balance

The fundamental principle enunciated by Keith Royer and Bill Nelson was arrived at after a considerable amount of research. They experimented with varying flows in relation to the flow rate determined by the Iowa formula. This flow rate they called the ideal rate of flow. So what are the consequences of using too little water applied with less than the ideal rate of flow?

For example, suppose that a 10,000ft^3 (283m^3) room is involved in a fire. By the Iowa formula,

$$\text{NFF (30}_\text{S}) = \frac{10,000}{100} = 100\text{gpm}$$

This means that 50gal (189L) is the right amount of water to control this size of fire. Suppose further that the available fire flow (AFF) is only 10gpm (37.8Lpm) and that the fire is well ventilated. At the rate of flow, it would take 5min to apply the needed 50gal (189L). Five minutes is certainly too long a time to achieve fire control. So what are the consequences of using too little water with less than the ideal rate of flow? Keith Royer has stated this about such an attack:

When the rate of flow is too small in relation to heat production and accumulation, there may be only a slight temporary reduction in temperatures in the fire area. The rate of heat production in remote areas may be increased.[7]

Figure 2-2 is a composite (or average) of a number of experimental fires from the research done at Iowa State University in which too little water was used. The solid line gives the temperature for the fire area, and the dashed line gives the temperature for an adjacent room or hallway and an upstairs room or an attic space.

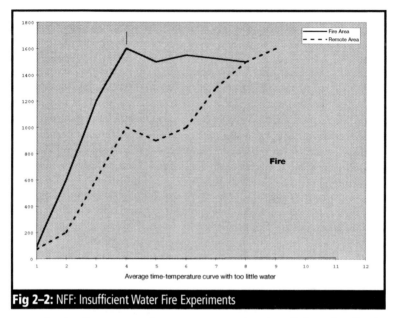

Average time-temperature curve with too little water

Fig 2–2: NFF: Insufficient Water Fire Experiments

In the fire room, the temperature levels off, but the fire is not brought under control. In the remote area, there is little effect upon the temperature. All of this is to be expected since there will not be enough water to spread unvaporized drops throughout the fire area. Truly, too little water will have little effect upon a fire.

What if too much water is used? In the previous example with an NFF of 100gpm (378.5Lpm), suppose that a flow rate of 250gpm (946Lpm) is used. At this rate for 30s, 125gal (473L) of water is applied to the fire. Yet only 50gal (189L) is actually needed to fill the confined space full of steam. Surely the more water you can put on a fire, the faster fire control can be achieved. This is a common belief, but it is not true for confined fires—it is a myth.

One of the great contributions of Nelson and Royer is their discovery that applying too much water to a confined fire is counterproductive. Figure 2-3 shows what happens in this case. Note the rapid fluctuations, or turbulence, of the temperatures both in the fire room and in the adjacent areas.

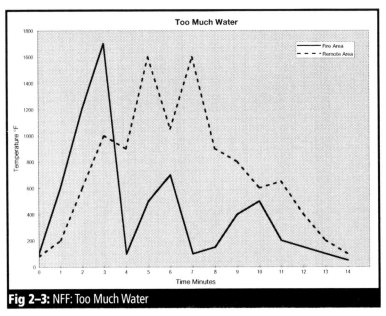

Too Much Water

— Fire Area
- - - Remote Area

Temperature °F

Time Minutes

Fig 2–3: NFF: Too Much Water

Royer and Nelson clearly explain why this occurs.

- If the ideal rate of flow is used, thermal balance is pre-served.

- If too much water is used, thermal balance is not pre-served.

Preserving thermal balance is what brings quick control of a fire. Destroying thermal balance (thermal imbalance) disrupts and delays control. Royer discusses the bad consequences of creating thermal imbalance:

> When the Rate-of-Flow is substantially greater than ideal, the temperature drop in the fire area will be very rapid, steam production will be limited, and heat production will increase in remote areas. Smoke from smoldering fuels will hang in the cool atmosphere of the area, complicating over-haul, sometimes to the extent that spot fires will come back. This will also delay control of fires in the remote areas, especially if the only approach to these areas is through the areas of fire involvement.[8]

Creating thermal imbalance could very well "push" the fire by allowing very hot gases to escape the cooling process and migrate to adjacent areas.

Key Test Fire

Thermal imbalance is created if too much water is used or if water is not distributed properly—that is, too much water in parts of the area. All of this was discovered in a test fire held in Des Moines, Iowa, on November 14, 1958. One purpose of the test was to determine whether the Iowa formula worked for a fire in a large building. The two-story brick building had wood joist construction and was approximately 55ft x 120ft x 10ft high on each floor (66,000ft^3 or 1,868m^3). By the Iowa Rate-of-Flow formula,

$$\text{NFF (30}_\text{s}) = \frac{55 \times 120 \times 10}{100} = 100\text{gpm (2,271}_\text{Lpm})$$

The attack was made on the completely open 2nd floor. Eight 1.5in (38mm) lines were used to control the fire. One unexpected result was that good overhauling conditions were not created in this test fire mainly because of uneven distribution of water. Too much water was applied along the edge of the area, and insufficient water was applied in the central sections. This application left hot spots down the center of the area and caused turbulent circulation of steam and smoke. This turbulent circulation that effectively delayed overhauling for 18min might be referred to as thermal or convection imbalance.[9]

Royer added that poor application created a thermal trap that resulted in thermal imbalance, and this just upset everything. The final report on this test fire contains the following conclusion:

> The importance of maintaining thermal balance in the area during the entire extinguishing process is vitally important. Without thermal balance overhauling will be delayed. Thermal decomposition and even some open burning will continue behind clouds of steam and smoke.[10]

Now you understand why Bill Nelson created two final fire stages: Post-Attack Warm (7A) and Post-Attack Cool (7B). It does make a difference how well, or how poorly, you handle a fog nozzle.

The data compiled by Nelson and Royer confirm that, with the proper rate of flow and proper distribution, no fire is "pushed" anywhere. In fact, a properly executed attack will indirectly control or extinguish fire in remote areas. In reality, there is no problem because there is no longer any fire to push.

Royer defines thermal balance in this way:

The amount of air that comes into the fuel is directly proportional to the amount of combustion that goes away from the fuel. So basically what a fire is trying to do...it is trying to seek equilibrium between products getting away and products coming in. Anything that disrupts one or the other, it throws it out of balance.[11]

This happens because of the law of conservation of matter and energy. The same amount of grams of substances consumed by the fire must equal the same number of grams of the products of combustion. Likewise, the energy content of the substances coming in and going out must be the same.

Another aspect of thermal balance is that temperatures are distributed uniformly horizontally throughout the fire area. Vertically, temperatures continuously increase from bottom to top, with the greatest concentration of heat at the highest level. In the early stages of a fire, if there are no openings above, temperatures tend to level off. This process is facilitated by the condensation of water vapor that comes in contact with walls and ceilings whose temperature is below 212°F (100°C). Water vapor condenses and releases the 971Btu/lb (2,262J/g) of heat absorbed when liquid water is vaporized to steam.

If a fire attack is made with the ideal Rate-of-Flow applying the right amount of water, then fire control is achieved in less than 30s. The following graph shows the steady decline of temperatures both in the fire room and in the adjacent areas.

Suppose a fire attack is made with the ideal rate of flow. The injection of little drops of water creates a violent disturbance that completely disrupts the thermal balance. The chemical chain reaction is stopped. Steam instantly absorbs the excess heat and fills the entire fire area. The steam literally blasts the fire out of existence with a powerful smothering force that deprives the fire of needed oxygen. Within a few seconds after the attack, steam begins to condense and appears as a white cloud that drifts upward and out of the structure. In other words, a cooler thermal balance replaces a hot thermal balance. Royer describes this situation as follows:

> If at the conclusion of the knockdown, the fire area is left with an even ceiling temperature of 300°F (148°C), conditions will be ideal for natural ventilation and easy and efficient overhaul. The lifting forces of the warm air (thermals) will be in balance throughout the area and we can say that we have left the area with the same thermal balance that was developed as the fire built-up but at a somewhat lower temperature.[12]

The discovery of the importance of maintaining thermal balance and the bad things that happen when thermal imbalance is created are some of the most important results of the research done at Iowa State University. The crucial discovery is that thermal imbalance and "pushing a fire around" happen only when fog nozzles are misused.

It is appropriate to end this section with a statement by Keith Royer that describes a truly effective fog attack:

> When the Rate-of-Flow is ideal, results will be best. Distribution is proper and the flow is stopped when the fire blacks out (usually in 20s to 30s). The area will be filled with steam and temperatures will continue to fall after application is stopped. Steam will flow into remote areas where heat and fire may have spread temporarily and suspend the production in these areas. Temperatures in the area of involvement will be relatively high—200°F to 400°F at ceiling level—and smoke in the area will lift rapidly. Overhaul crews will be able to move in to extinguish spot fires in both the main area and the remote area.[11]

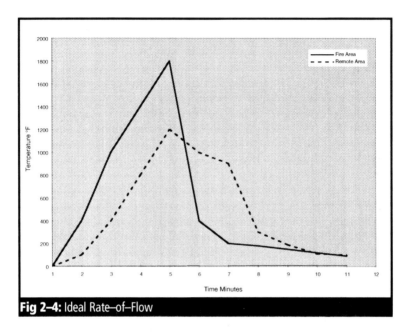

Fig 2–4: Ideal Rate–of–Flow

The Combination Attack

In looking back to the time when Bill Nelson and Keith Royer made their contribution, it might seem that the research was relatively easy. On the contrary, the research required several years of effort and many experimental fires before any conclusions could be made. The research was complicated by the need to bring together three things:

- The right amount of water

- The right tool (nozzle)

- The right way to apply water

The gallonage formula solved the problem of determining the right amount of water. To address the second issue, Royer and Nelson participated with the Akron Brass Company in developing a new type of fog nozzle, the constant flow nozzle. Earlier fog nozzles varied flow as the fog pattern was adjusted. The constant flow nozzle, on the other hand, maintained the same flow as the pattern was adjusted from a straight stream to a wide-angle fog pattern. This was accomplished by using a sliding sleeve in the nozzle. Obviously, this constant flow nozzle was needed because of the importance of applying the right amount of water to a fire.

Royer and Nelson also participated in the development of large-volume fog nozzles. The Iowa formula told them that for larger buildings, hand lines could not provide adequate fire flows. So this led to the development of snorkels and other elevated master stream devices.

With the right tools available, the next question that confronted Nelson and Royer was, What was the right way to use the constant flow nozzle? After trying various ways to apply water, in 1959 they finally tried a new way to distribute by rotating the fog nozzle just inside a window or doorway. This was the birth of the combination attack, which Royer describes here:

> The proper distribution of water fog on the initial application is very important, and the nozzle operator must understand what result is trying to be accomplished. The fog pattern is adjusted so it will just reach across the area and rotated following the contour of the area—striking as much of the perimeter of the area as possible, with the outer surface of the fog stream across the floor, up the side, across the ceiling, etc. This rotation is as violent as it is possible for the nozzle operator to make it, in placing the nozzle inside the area, it should be inside the window or other opening about an arm's length.[14]

They strongly recommend that the nozzle rotation be clockwise (from the nozzle operator's position). After many experiments they determined that a clockwise rotation is more effective and safer than a counterclockwise rotation.

- Clockwise rotation drives smoke, gases, and flames away from the nozzle. Counterclockwise rotation does just the opposite.

- Clockwise rotation produces steam with an active rolling action. Counterclockwise rotation produces steam with an inactive and lazy action.

- Clockwise rotation produces a faster knockdown time.

These differences may be related to the production of ions in a fire or may simply be because clockwise rotation is smoother for right-handed individuals. But the true reason is not conclusively known even today.

Iowa State Research Summary

The research done by Keith Royer and Bill Nelson provided additional major contributions to the safe and effective use of fog nozzles.

1. The gallonage formula determines the ideal rate of flow for a confined fire of a given size.

2. The Iowa formula determines an ideal rate of flow that is highly useful in preplanning to measure the capability of a fire department to provide the NFF for the largest open area of a given structure.

3. The constant flow nozzle is the right tool to flow the right amount of water in a fog attack.

4. The combination attack is the most effective fog attack with an efficiency of better than 90%.

5. Thermal imbalance is created by using too much water, which emphasizes the importance of applying the right amount of water in a fog attack.

Layman's research, taken all together with Royer and Nelson's research, provides substantial additional information needed for a safe and effective fog attack. For example, there is widespread controversy today whether one method of fire attack is better than another method. This particularly applies to the indirect method versus the direct method. Many disparaging statements have been made about the indirect method, advocating that it not be used. Others say that the direct method is the only method that should be used. Actually, both are wrong.

The most effective fog attack is a combination of the indirect method and the direct method—in other words, the combination attack. With a better than 90% efficiency, the combination attack provides the most effective fog attack with the least amount of water in the shortest period of time. So why is there any controversy about which method to use? The simple answer is to use them both for a safe and effective fog attack.

The contributions of Keith Royer and Bill Nelson can be stated by adding to the fundamental principle the following phrases:

> In principle, firefighting is very simple. All one needs to do is put the right amount water in the right place using the right tool in the right way.

This briefly summarizes the contributions that came from their research at Iowa State University. Unfortunately, their research was not well known throughout the United States. Consequently, many fire departments did not have adequate guidance, based upon scientific research, to guide them in the safe and effective use of fog nozzles.

Now let's turn from the research at Iowa State University to the experience gained elsewhere in the use of fog nozzles. The contrast between the successful research at Iowa State and the failure to learn how to use fog nozzles safely and effectively is quite startling.

The Identity Error

During the 1950s and 1960s, a great deal of experience was gained in the use of fog nozzles. However, the only part of the research at Iowa State University that received nationwide attention was the Iowa Rate-of-Flow formula. Nelson and Royer published several articles in Fire Engineering magazine, but Bill Nelson's book, *Qualitative Fire Behavior*, did not appear until 1991. The combination attack did find its way into the *Essentials of Fire Fighting*, but there was little guidance on how to use it safely and effectively. So, much of the experience with fog nozzles was gained without the benefit of any research.

Many departments went wrong in attempting to apply the Rate-of-Flow formula:

$$\text{NFF} \times t = \text{gal} \qquad \text{NFF} \times t = \text{L}$$

It is clear that t (time) is an essential element of this formula and cannot be eliminated. The equations

$$\text{NFF} = \text{gal} \qquad \text{NFF} = \text{L}$$

are true only if $t = 1$:

$$\text{NFF} \times 1 = \text{gal} \qquad \text{NFF} \times 1 = \text{L}$$

In other words, if NFF = 100gpm (Lpm), then gal = 100 (L = 100) if and only if $t = 1$. For every other time this equation is false.

This is what happened to the Iowa Rate-of-Flow formula:

$$\text{NFF} = \frac{\text{Vol}}{100} \qquad \text{NFF} = \frac{\text{Vol}}{.75} \ (\text{metric})$$

This equation is true only if $t = 1$ (minute):

$$\text{NFF} \times 1 = \frac{\text{Vol}}{100} \qquad \text{NFF} \times 1 = \frac{\text{Vol}}{.75} \ (\text{metric})$$

But the Iowa formula is not valid for 1min; it is valid only for 30s. This constitutes a misuse of the Iowa formula that results in applying twice as much water on a given size of fire. Applying twice the right amount of water causes thermal imbalance and was the source of many of the problems that fire departments experienced during this time.

So the mathematical error, the identity error, was the first of several problems that fire departments encountered in trying to learn how to use fog nozzles safely and effectively.

The Shut-off Valve

The constant-flow nozzle solved one problem, that is, the nozzle flowed the right amount of water no matter what changes were made in the fog pattern. However, an equally serious problem remained to be solved.

Most of the nozzles produced during this time had a ball shut-off valve. In effect this type of nozzle had only two flow rates: zero when shut off and maximum when wide open. Any attempt to partially open the ball valve resulted in extreme turbulence in the nozzle with significant deterioration of the fire stream. Further, there were no in-between settings to indicate how much reduction in flow was obtained. The problem worsened with the introduction of 1.75in and 2in attack lines. This resulted in increases in maximum flows from 150gpm to 200gpm (577Lpm to 757Lpm). The flow rate from these lines far exceeded the flow rate needed to control room-size fires without creating massive thermal imbalance problems

Remember that Royer and Nelson determined that flow rates near the ideal rate of flow were needed, requiring nozzles with the capability of varying the rate of flow. The ideal rate of flow increases as the volume of fire increases, and the ideal rate of flow for room-size fires is much less than 100gpm (378Lpm)—in the range of 30gpm to 60gpm (114Lpm to 228Lpm). This rate of flow is well below the standard flows from fog nozzles attached to 1.5in (38mm) or larger attack lines. So with the wrong type of shut-off valve, fire departments began to violate the fundamental requirements for the safe and effective use of fog nozzles.

Gross misuse of fog nozzles happened in these ways.

- Instead of varying the flow rate less than 100gpm (378Lpm) for room-size fires, ball nozzles had one flow rate—wide open—that applied too much water for smaller fires.

- Instead of distributing the water evenly throughout the fire, water was concentrated in one part of the fire area.

- Instead of shutting down the nozzle in seconds at the proper time, water flowed for minutes—far longer than needed.

In fact, much of the research that was available then was of no help to fire departments in using fog nozzles. Bill Nelson complained, in Qualitative Fire Behavior, that many of the tests sponsored by the Exploratory Committee were not genuine tests at all. He added:

> In retrospect, some dumb things were done during many of these tests. Some thought that the angle of the fog was very important for the successful indirect application, and the fog streams were very carefully calibrated at a 30° angle with large wooden calipers. This particular feature of fog use was so blown out of proportion that some fire chiefs actually had their fog nozzles locked at a given set angle.[15]

He also said that, in general, more was learned when carefully staged demonstrations went wrong. He summarized the situation at that time in this way:

> At any rate, tests were run and discussions and arguments were often heated during the 1950s. About 1960, the pendulum swung. Fog nozzles became fashionable for all fires in most fire departments and the fire service was plunged into a dark decade of thermal imbalance. Buildings continued to burn while the fog streams flowed. Slowly firefighters learned that fog was a useful tool, but needed to be carefully used if it is to save more buildings than it lost. Certain stubborn individuals continued to point out that straight streams used properly were just as effective as fog and usually do not cause thermal imbalance problems.[16]

Bill Nelson and Keith Royer certainly counted themselves among those stubborn individuals. Quite simply, the situation that existed then was that the available research was not adequate to guide all fire departments in the proper use of fog nozzles. Unfortunately, trial-and-error methods do not guarantee that anyone will learn how to use fog nozzles safely and effectively. That is evident from the statements that Bill Nelson made.

"Pushing a Fire"

So without adequate guidance from research, fire departments continued to misuse fog nozzles, thereby producing massive thermal imbalance problems. One aspect of thermal imbalance that dominated thinking throughout the United States was pushing a fire. The admonition was heard frequently and seen often in print: "Don't push a fire around."

A perfect example of pushing a fire is presented in William E. Clark's, *Firefighting Principles and Practices.*[17] Clark's first picture of a house shows a corner room with fire coming out one window where fog was applied. Black smoke is coming out the other window around the corner. Fog is being applied in a straight stream through the fire window by a firefighter standing some distance from this window. A second picture shows that while fire was blanked out at the window where fog was applied, fire erupted from the second window around the corner. The same thing happened at the inside door—fire spread through this door. Clark commented that the fire was driven throughout the entire 1st floor. Because water was projected through the window in a narrow stream, the fire was not controlled. Further, pressure was created to blow the fire out the second window and to spread fire into adjacent areas of the house. This is a classic case of misuse of a fog nozzle that results in fire spread, or pushing a fire.

Another example with tragic consequences happened in a multistory residential building. One apartment on the top floor was on fire with its hallway door open. Several civilians and firefighters were on the fire floor at this time. The officer–in–command ordered an aerial ladder with a master stream nozzle to hit the outside apartment window. He ordered the nozzle used for 30s. Assuming the nozzle flowed 500gpm (1,892Lpm), then 250gal (946L) were applied into the apartment. Assuming further that the apartment was no more than 10,000ft^3 (283m^3), the right amount of water for that size fire is 50gal (10,000/200 = 50) (189L).

Thus, five times the right amount of water was applied to that apartment, with the inevitable and predictable result. A blast of smoke, heat, and gases filled the hallway. Some civilians and firefighters escaped with their lives, but others did not. You can call this pushing a fire if you wish. However, blasting a fire is a more accurate description of what happened.

The phrase "pushing a fire" is somewhat ambiguous. Fire is a chemical reaction in which fuel gases unite with oxygen (a gas) to produce other gases and carbon (a solid) with the release of heat and light. Gases can be pushed (or moved) only by pressure or gravity. So projecting a fog stream into a confined space does not necessarily increase the pressure in that space. If a narrow fog pattern is used, what is more likely is that convention currents will be created around this stream. This creates turbulence and does not necessarily increase the pressure in the space.

It is a little strange to think about pushing a fire with water. The question is, Precisely how can this be done? After all, liquid water applied to a fire is vaporized to steam at 212°F (100°C). This sudden blast of steam can absorb all of the excess heat produced by the fire and stop the combustion process cold. Hence, there is no more fire to push anywhere. Further, "pushing a fire" seems to say that somehow you can take a fire and push, or move, it intact into an adjacent area while it is still burning and it will continue to burn there. That is difficult to imagine. A more likely scenario is that fire gases are pressurized, or forced, into adjacent areas, and the heat vaporizes the fuels there. If enough oxygen is present, then ignition occurs. This can only happen if the fuel in the adjacent area has been preheated from adjacent fire production. If pressure or turbulence cause hot gases to migrate, these surfaces can be brought to ignition temperatures quickly, leading to the illusion that fire has magically been pushed or transported.

So this simple–sounding phrase, "pushing a fire", turns out to be rather complex in nature, and you cannot be certain at all about its exact meaning. At any rate, the only rational explanation comes from the research at Iowa State University. Royer and Nelson discovered, the hard way, the bad consequences of creating thermal imbalance. One consequence of thermal imbalance is pushing or spreading the fire into adjacent areas. However, thermal imbalance happens only if too much water is used or if too much water is used in parts of the fire area (poor distribution). Using the right amount of water does not create thermal imbalance, and there is no problem with pushing a fire nor threat of anything else bad happening. What happens with the right amount of water is that steam expansion pressurizes the space and displaces the contaminated atmosphere. This expansion carries unvaporized particles of water into adjacent areas where they are in turn vaporized into more steam. This is the indirect effect of a fog attack that controls a fire in areas adjacent to where water was applied.

Turnaround Tactics

Without the benefit of the research done at Iowa State University, many fire departments hit upon a simple but wrong solution to the problem of pushing a fire. If a fog nozzle operated from outside a structure pushes fire farther inside, why not turn things around and attack from the inside and push the fire outside? Thus was born the strategy, "Always attack a fire from the unburned side of the structure." The problem with this strategy is the word *always*. Two examples will demonstrate quite easily that this strategy is not always a good idea.

The first fire call that the Kittrell Volunteer fire department in Tennessee responded to after it was created in 1991 was a fire in a one–story house with an attached room on the rear with a shed roof and a single door connecting to the rest of the house. This room was well involved in fire when the department arrived on the scene. The chief decided to make a direct attack on the fire using two 1.5in (38mm) lines. At this time, a second crew arrived and proceeded through the front door, going around furniture, into a short hallway, and then into a room at the rear but still in the main part of the house. They had to find the door leading to the add–on room that was on fire. By the time this crew reached the rear door, the outside direct attack had extinguished the fire. While all of this was going on, a crew had laddered the peaked roof over the main part of the house, chopped a hole in the roof at the peak, and extinguished a small fire in the attic. If our department had followed the admonition, "Always attack from the unburned side," the delay inherent in an inside attack would have resulted in a serious attic fire above the main part of the house with subsequent danger to firefighters.

Another fire the same year occurred in a vacant house. The fire was burning in a middle bedroom whose window was on the right, next to the front door. The chief removed an air conditioner attached to the lower half of the window, and a combination attack was made through this hole. The fire was blacked out immediately. Just at this moment an officer from a mutual aid department ran up and yelled, "Don't push the fire!" It was too late to bring the fire back. Not believing that the fire could possibly be extinguished, the officer sent his crew through the front door with a charged line. The crew had to detour to the left because of a divider straight ahead of the front door. At this moment, John Wiseman decided to circle the building. When the author arrived at the back door, there was the mutual aid crew looking for the bedroom, confused and disoriented by smoke and the house layout. John asked, "What are you guys doing here?" He informed them that the bedroom was on the other side of the house in the front. This wasn't a fatal mistake, but no crew needs to be unnecessarily wandering inside a burning structure. The faster you can locate a fire and attack it, the better off you are. It is not necessary to always attack a fire from the unburned side.

Misinformation

The research done by Layman, Royer, and Nelson was not well known throughout the United States, and much of what was published over the years was misinformation. Here are some examples of misinformation that was published.

Chief Richard A. Knopf, of the Portage, Michigan, Fire Department, wrote an article for *Fire Chief* in 1979. Knopf complained that we, the recipients of Chief Layman's techniques, "...have expanded them far beyond their intended application to the detriment of the overall extinguishment process."[18]. Knopf quoted Layman's statement that degree of confinement and concentration of excessive heat is an important factor in the indirect method of attack. Somehow, Knopf interpreted this to mean a stage three hot smoldering fire. Since only a limited number of fires enter stage three, Knopf concluded that the indirect method of attack was of very limited usefulness. In other words, departments had expanded the usefulness of the indirect method beyond the usefulness intended by Layman.

However, Layman's six case histories documented Class A structure fires. All of these cases were common stage two fires, and not a single one was a hot smoldering fire. So it was Layman who expanded the indirect method far beyond stage three hot smoldering fires, not misapprehension by others.

Knopf continues: "Many of Lloyd Layman's experiments were conducted on board a ship."[19] He concludes that Layman's work must be kept in proper perspective and that the indirect attack must be used under the proper conditions. Does this mean onboard ships only?

Knopf presents a third argument that the primary focus of fire suppression should be on the interior attack directed at the center of the fire. Wisely, Knopf does not say "the exclusive focus". He recognizes that an interior attack is not always possible for fires in the third stage or near flashover. He continues, "In most other cases, an aggressive interior attack is the most effective way to minimize loss."[20] Unknowingly, Knopf is stating exactly what Layman advocated. Layman said, with respect to stage one fires:

A fire of this nature must be located and extinguished by direct attack. In making the initial size–up of a building fire, an experienced and capable officer should have little difficulty in determining if the situation demands a direct or an indirect attack.[21]

The focus should not be determined by the frequency with which a given method of attack is used. All methods have their proper place and, when used properly, are more effective than any other method for that particular fire.

Another critic, Harold Richman, retired fire chief from Reston, Virginia, recommends solid–stream nozzles, or fog nozzles adjusted to the "solid–stream" setting for the most effective interior fire attack. He states, "Use of fog streams inside a building should be restricted to unoccupied confined spaces such as attics."[22] He adds, "Only with adequate ventilation ahead of the nozzle should a fog pattern be used inside a building."[23] His reasoning is based upon the fear that fire, smoke, gases, and steam could roll over and around firefighters.

These two statements by Richman are contradictory. If adequate ventilation ahead of a fog nozzle can be provided, then there is no need to prohibit the use of fog nozzles in confined spaces other than attics. Even if a confined space is occupied by firefighters, apparently that is okay provided ventilation is adequate. So Richman does not provide a complete explanation for his recommendation. It is fair to ask whether he has research to support his conclusions.

It would be rather curious, indeed, to restrict the use of fog streams to attic spaces. To say the least, this recommendation is unique, and we have not encountered a similar recommendation from anyone else, whatever the reasoning. Fighting an attic fire from underneath does not eliminate all of the dangers that confront firefighters. An attic attack from underneath, as Richman recommends, places firefighters in danger even though they do not occupy the attic space. Collapse of ceiling (sheetrock) or ceiling joists presents a real danger. In fact, the only serious injuries that have occurred within the past two years in the county in which the authors live have occurred to firefighters while using this tactic.

A third naysayer, Dave Clark, an instructor at the Illinois Fire Service Institute, makes this statement:

As soon as a fog stream is opened, bad things start to happen.

- You lose visibility.

- Air is entrained and flames increase in size.

- Heat and smoke is pushed ahead of the fog stream.

- All this moves back toward and over the hose crew.[24]

He then adds, "Keep in mind that the idea is to put the water on the fire, and to not generate steam. Anything that you do that makes steam works against you."[25] These are truly amazing statements that indicate a complete lack of understanding of water behavior in a fire.

The first statement is a perfect example of misuse of a fog nozzle that creates thermal imbalance. Using the right amount of water does not produce any of these bad things. Learning this lesson is the key to understanding how to use fog nozzles safely and effectively.

The second statement is that it is possible to put water on a fire and not generate steam. This statement is not qualified in any way, so that means not to generate any steam at all. Clark's statement is absolutely impossible and contradicts all scientific facts about water behavior in a fire.

- Liquid water projected in a fire changes to a gas (steam) at 212°F (100°C).

- Ignition temperatures of all hydrocarbon fuels are well above 212°F and usually about 400°F to 500°F (204°C to 260°C).

- Fire temperatures are much hotter, ranging from 1,100°F (593°C) to a maximum around 2,000°F (1093°C).

- Thus, liquid water is instantly converted to steam in a fire at 212°F (100°C) and the steam temperature does not rise above 212°F (100°C) this physical change of state is endothermic, that is, heat absorbing.

- At the same time, the formation of steam results in an expansion of 1,700:1 or $1ft^3$ (28.3L) of liquid water forms $1,700ft^3$ (48,100L) of steam.

What these scientific facts about water behavior prove is that it is impossible to put water on a fire without generating steam. In other words, it is impossible for water to remain liquid in a fire for long. The only question is, how much steam is created? If the efficiency of the conversion process is low, not all of the liquid water applied to a fire will be converted to steam. However, the methods developed at Iowa State University have an efficiency of better than 90%.

Clark's statement that anything you do that makes steam works against you couldn't be further from the truth. In fact, we would be severely handicapped if water did not turn to steam at 212°F (100°C). Of the 9,330Btu absorbed by 1gal of water, more than 8,000Btu are absorbed when water is vaporized to steam. Without this firefighting power, we would be almost helpless in fire combat. The heat–absorbing capacity when water is vaporized to steam (called the enthalpy of vaporization of water) is far greater than that of any other liquid which could possibly be used to fight fires. With water turned to steam, we can fight fire on equal terms; without it, we would face a losing battle most of the time.

Instead of saying that steam works against you, in truth the key to effective firefighting is to make steam work for you. In fact, the safest and most effective fire attack is a fog attack that applies the right amount of water to the right place in the right way with the right tool.

Clark addressed another issue in the following statement:

For Layman's situation of confined fires in steel ships with no life hazard, this procedure works. It was never intended for interior attack on structural fires in which we have open areas, void, and people present, both firefighters and civilians.[25]

Obviously, Clark has not read Layman's book and is not aware that Layman successfully adapted the indirect method of attack to Class A structure firefighting. This is another misconception that can be easily disproved. Layman stated that:

In answer to the question regarding the effect on occupants of steam from fog application, we can only state that we have never heard of any adverse effects. Contrariwise, the much more rapid flame suppression with indirect application makes it possible to reach endangered persons more quickly so as to be able to remove them to safety and render aid as necessary.[26]

Bill Nelson also addressed this issue with the following:

Some individuals have warned that it is dangerous to use fog to fight fires in rooms of structures where people may still be trapped in other parts of the building. This is certainly true where the bulldozer–attack might be used or where fog lines are used that are larger than necessary for a particular open area. Such lines will push a considerable amount of products of combustion, steam, and fresh air ahead of them. However, if proper size lines are used and handled in the proper manner, the products of combustion and steam forced into remote areas will be cooler than those originally being forced there by the fire. In other words, immediate application of proper rates of flow of water can help occupants trapped in the building rather than endangering them.[28]

This is a perfect response to Clark's allegation.

Clark also made the following observation about firefighting in the 1950s and later: "The technique [exterior attack] had limited merit in the years prior to widespread use of breathing apparatus and Nomex turnout gear when most departments were restricted to exterior operations."[29] The research conducted by Layman, Royer, and Nelson was done with firefighters using air masks and making interior attacks. Royer, when asked about this, said they did quite a bit of interior firefighting. It is true that some departments did not have air packs, but this did not restrict them to exterior firefighting. They routinely went inside a burning structure. In other words, they were true smoke eaters.

However, the same Scott air packs were used then as now. The only difference was that the packs were carried in suitcases stored in a compartment on the fire truck. The turnout gear was adequate for interior firefighting. The main difference was that rubber boots were worn pulled up to the hips.

Finally Gene Carlson concludes with the following declaration:

Today Chief Layman's theory is dead.[30]

What is Chief Layman's theory? He was interested in explaining how the indirect method of attack worked. Since no water was applied directly on the burning fuels located beneath the deck plates of the engine room, exactly how could the fire be extinguished? Layman stated his theory this way:

The injection of water into a highly heated atmosphere results in rapid generation of steam, thereby, creating an atmospheric disturbance of sufficient force to distribute unvaporized particles throughout the space. Unvaporized particles are brought into contact with heated materials located beyond the immediate area, thereby exerting cooling action throughout the atmospheric area, and at the same time contributing to the atmospheric disturbance by expanding into steam.[31]

Carlson, presumably, is not challenging the two scientific facts that underlie Layman's theory:

1. Liquid water is transformed into steam at 212°F (100°C).

2. The expansion ratio of liquid water to steam is 1,700:1 at 212°F and the temperature of the steam is not raised above 212°F.

Layman theorized that unvaporized particles are blasted throughout the confined space, and these particles in turn are vaporized by contacting heated materials. This explains how fire can be extinguished remote from where the water was directly applied. Is Carlson claiming that a direct attack does not work? It would be nice if Carlson had some experiments, or research, to back up his declaration. Until then, we think that it is safe to say that Carlson is mistaken in his conclusions.

What is Chief Layman's theory? Chief Layman describes how his indirect method works as follows:

> If the volume of steam generated within the space exceeds the net atmospheric volume of the space, most, if not all, of the original atmosphere will be displaced by steam. When the surface temperature within the space has been reduced to approximately 212°F (100°C), the boiling point of water, steam generation ceases. At this time, steam within the space starts to condense and cold air from the outside enters filling the void created by the process of condensation. This in draft of cool air from the outside atmosphere tends to increase the rate of condensation and continues until the process of condensation ceases. At this time, a major part, if not all, of the atmosphere within the space consists of normal air.[32]

This clearly explains how the indirect method of attack with atmospheric displacement works. It worked when Layman first created this new method of attack. It works today, and it will work in the future.

All of these misstatements have an underlying theme: Fog nozzles cannot be safely or effectively used under any circumstance. Yet the research conducted from the 1940s until the 1980s points to the opposite conclusion. It would be strange indeed if fog nozzles could not be safely and effectively used because of some fatal flaw. Let's now turn to more recent research that will add further proof that fog nozzles can be used safely and effectively.

Notes

[1] Floyd W. (Bill) Nelson, *Qualitative Fire Behavior* (Ashland, MA: International Society of Fire Service Instructors (ISFSI), 1991), p. 63 ff.

[2] Ibid, p. 62.

[3] Ibid, p. 102.

[4] Lloyd Layman, *Attacking and Extinguishing Interior Fires* (Quincy, MA: NFPA, 1955), p. 20.

[5] Keith Royer, "Water for Fire Fighting", Iowa State University Engineering Extension Service Bulletin 18, (undated) p. 1.

[6] John D. Wiseman, *The Iowa State Story* (Stillwater, OK, Fire Protection Publications (FPP), 1998) p. 28.

[7] Keith Royer, "Water for Fire Fighting", op. cit., p. 3.

[8] Ibid.

[9] Keith Royer, "Test Fire for Exploratory Committee on Application of Water", Iowa State University Engineering Extension Service (no number) (1959), p. 15.

[10] Ibid.

[11] John D. Wiseman, p. 30.

[12] Keith Royer, "Test Fire for Exploratory Committee on Application of Water", op. cit., p. 22.

[13] Keith Royer, "Water for Fire Fighting", op. cit., p. 4.

[14] Keith Royer, "Report of Story City Fire Test", Iowa State University Engineering Extension Service Bulletin (no number) (1959), (entire booklet).

[15] Floyd W. (Bill) Nelson, op. cit., p. 100.

[16] Ibid.

[17]William F. Clark, *Firefighting Principles and Practices* (Saddle Brook, New Jersey: Fire Engineering, 1991) p.. 36.

[18]Richard A. Knopf, "Fog Compared with Straight Streams", *Fire Chief* (Vol 23 No 7 July 1979), p. 36.

[19]Ibid.

[20]Ibid.

[21]Lloyd Layman, op. cit.

[22]Harold Richman, "Improving Interior Fire Attack", *Fire Command*, (Vol 53 No 7 July 1986), p. 18.

[23]Ibid.

[24]David Clark, "Let's Get Something Straight", Illinois FSI newsletter (Summer 1990) p. 4.

[25]Ibid, p. 5.

[26]David Clark, "Straight Talk about Nozzles and Fire Attack", unpublished newsletter Illinois Fire Service Institute, p. 4.

[27]Lloyd Layman, op. cit., p. 146.

[28]Floyd W. (Bill) Nelson, op. cit., p. 109.

[29]David Clark, "Straight Talk about Nozzles and Fire Attack", op. cit., p. 4.

[30]Gene Carlson, "Lloyd Layman's Theory: Its Time has Come and Gone", *Fire Engineering* Vol 136 No 2, Feb. 1983)

[31]Lloyd Layman, op. cit., p. 36.

[32]Ibid, p. 37.

CHAPTER 3

US Navy Research Laboratory

The research at Iowa State University continued for 28 years, beginning in 1951. The next major research effort did not begin until about ten years later. As a result of the missile attack upon the *USS Stark* in the Persian Gulf in 1990, the US Navy initiated a series of experiments in the 1990s to learn how to handle a fire resulting from burning missile propellant. To broaden the research, the researchers also decided to design an equivalent fire that would approximate the worst–case post flashover compartment fire. The designed fire was divided into two phases. In the first phase, the burning of the missile propellant would almost immediately produce flashover. In the second phase, there would be a transition to a steady–state post flashover fire involving fuel and Class A materials. In other words, there would be an initiating event following by a conventional fire.

The first series of tests in 1991 used a small–scale mock–up of four cubical compartments. There were six objectives for these tests:

1. Bound and quantify the conditions required to achieve maximum compartment fire temperatures under natural ventilation conditions.

2. Gather baseline data on the effects of vent and fuel surface area on compartment fire temperatures.

3. Based upon the above knowledge, design fires for characterization of post flashover fire threats, which then could be used on the ex–*USS Shadwell* for larger scale testing.

4. Develop preliminary estimates of vertical and horizontal fire spread rates along with human/material tolerance limits based on heat transfer characteristics to adjacent compartments.

5. Develop an overall lessons list to give preliminary guidance to the fleet and aid in better understanding of the tests to be conducted on the ex–*USS Shadwell*.

6. Begin preparations for future tests of active and passive fire protection extinguishing systems.[1]

The researchers analyzed fire growth in the usual way, with flashover serving as the transition between the early growth stage and the fully developed stage. They used the ISO (International Standards Organization) definition of flashover as the rapid fuel surface involvement of all combustibles in a given space. The researchers stated two thermal characteristics necessary and sufficient for flashover to occur: [2]

1. Heat flux to the floor of 1.8Btu/ft²/s (20kW/m²)

2. Upper layer temperature exceeding 500°C (932°F)

The first characteristic is probably unfamiliar to firefighters, while the second certainly should be. Heat flux is the rate of heat energy transfer to a given surface, such as the floor of a compartment. If the heat flux stated in point one does not equal 20kW/m², then flashover cannot occur. Likewise, the upper layer temperature must exceed 500°C (932°F). In other words, both conditions must be present for flashover to occur.

Consequently, compartment fire growth is divided into three stages:

1. Pre–flashover—A growth stage where the average temperature (*T*) is low and the fire is localized to the vicinity of the origin. For a propellant–induced fire, this stage is very short—less than 2min.

2. Post–flashover—A fully developed stage where all combustible items in the compartment are involved and the threat to adjacent compartments is greatest.

3. Decay Stage where the severity of the fire decreases substantially as the combustibles are consumed.[3]

The researchers further analyzed fire behavior by three burning regimes, or phases. This is a different analysis from Bill Nelson's analysis of seven fire stages.

The first regime is ventilation controlled, that is, before flashover and when not enough oxygen is available to burn all of the fuel gases being released. If there is a vent opening, these unburned fuel vapors leave the compartment and act as a heat sink. This means that heat is removed and temperatures in the compartment are actually lowered.

The second regime is flashover, a narrow time period when the amount of fuel being vaporized equals the amount needed to consume all of the oxygen entering the compartment. This is a balanced situation that releases the maximum amount of energy, thus being the most severe of the three regimes. This is called stoichiometric burning. In regime three, the post flashover fire, excess air flows into the compartment, and the rate of heat release is limited by the fuel surface area available and the burning characteristics of the fuels involved. This is a fuel–controlled fire, and the excess air has the effect of cooling the compartment making this regime the least severe of the three burning regimes.

The naval researchers used a standard experimental program developed in the laboratories of eight European nations. These experiments varied the size and shape of the compartment, the size and height of the vents, and the fuel load. From this, a formula was derived that calculated a numerical value for an opening factor:

$$X = \frac{A_t}{A_w \times \sqrt{H}}$$

In this equation

A_t = total surface area minus the floor and vent areas (in square meters)

A_w = vent area (in square meters)

H = height to the top of the vent (in meters)

X = value for opening factor

Researchers determined that a ratio of total area to vent area of

$$8/\sqrt{H} \quad \text{to} \quad 15/\sqrt{H}$$

produced the maximum temperature for a compartment. For an average–size room, this number range corresponds to a vent area of one window and one door. Figure 3–1, known as the Thomas curve, shows the data used to reach this conclusion. Notice that the peak rate of heat release is in the middle of the curve that corresponds to the values of the formula from $8/\sqrt{H}$ to $15/\sqrt{H}$. It may come as a surprise to find out that more ventilation doesn't necessarily produce a hotter fire.

First test series

The first test series consisted of 60 test fires. JP–5 fuel oil was flowing into a pan on the floor at a rate of 7.6Lpm (2gpm). Test 27, with two doors (Navy type) open into the compartment

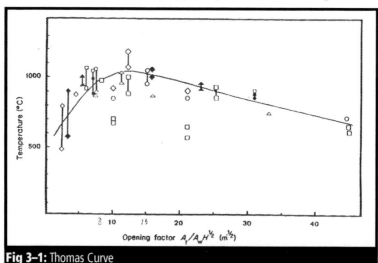

Fig 3–1: Thomas Curve

(opening factor = 9.6), produced the hottest fire. The upper layer temperatures reached 1,100°C (2,012°F), with the total heat flux to the floor of 50kW/m² to 120kW/m² (4.4Btu/ft²/s to 10.6Btu/ft²/s). These values were well beyond the values needed to produce flashover. With either one or three doors open, the upper layer temperatures and the average temperature were lower. These data confirmed the research done in the European labs.

Thus, the researchers were satisfied that they had produced the design fire that would be used in full–scale testing on the ex–*USS Shadwell*. The opening factor would be approximately 10/ H. The air flow would then be calculated, and that in turn would determine the fuel flow rate that would produce the most severe conditions in the fire compartment. We should note that a ventilation–controlled fire may produce a greater threat because of flames extending from the compartment passageways or into adjacent compartments through doors, hatches, or bulkhead breaches.

One objective of the testing was to determine the rate of fire spread into adjacent compartments. The bulkheads and decks, of course, were made of steel. With the high conductivity of steel, the temperature difference between the inside surface and the outside surface was very small, remaining constant at about 50°C (122°F). The temperatures on the outside surfaces exceeded 750°C (1,382°F). Thus, the adjacent compartment would quickly become untenable for humans not wearing protective clothing or air masks.

There were two criteria given in the report for determining human tenability. These values were minimal, or the threshold beyond which humans could not survive in a compartment:

1. Radiant heat flux from upper layer temperature of 183°C (361°F)

2. A lower level temperature of 100°C (212°F)

The radiant heat flux was based upon damage to bare skin. The lower level temperature was based upon lung damage from inhaling hot gases. Based upon this data, fire would spread to adjacent compartments, making them untenable within 2min. Fire would spread to adjacent compartments within 6min to 7min after flashover in the original compartment. This is the time it would take to reach fuel ignition temperatures in the adjacent compartment.

This test series established two very important principles:

- There is no direct linear correlation between ventilation and fire severity. In other words, more ventilation does not necessarily produce a hotter fire. The ventilation that produces the hottest fire is somewhere in the middle for the size of vents, with an opening factor of $8/\sqrt{H}$ to $15/\sqrt{H}$. With a smaller opening, or a larger opening, fire severity is less.

- Likewise, there is no direct linear correlation between fuel flow and fire severity. In other words, more fuel does not necessarily produce a hotter fire. The fuel flow that produces the maximum fire temperature and heat flux is somewhere in the middle—7.6Lpm (2gpm) for these test fires.

Second test series

The objective of a second series of tests was to determine the best methods for stopping horizontal and vertical spread of a compartment fire. Earlier research led to the belief that a very low flow rate was needed to stop the spread of such a fire. If $50kW/m^2$ ($4.5Btu/ft^2/s$) of heat is conducted through an aluminum deck, then it was theorized that it could be cooled to below 100°C (212°F) with an application rate of $1.34Lpm/m^2$ ($0.33gpm/ft^2$). This assumes 100% efficiency—that is, 100% conversion to steam. At lower efficiency rates, probably an application rate of $4.07Lpm/m^2$ ($0.1gpm/ft^2/s$) was needed. At this rate a fog nozzle flowing 95gpm (360Lpm) could cool a $93m^2$ ($1,000ft^2$) bulkhead or deck.

The most important conclusion from these tests was to establish the optimum application rate for boundary cooling to 100°C (212°F). The application rate above this level produces minimal additional cooling. The key rate is[5]

$$2.04Lpm/m^2 \ (0.05gpm/ft^2)$$

Applying greater rates resulted in pooling of water, leading to collecting enough water to cause the ship to list. Application of lower rates had a minimal effect on cooling boundary surfaces.

For horizontal cooling, applying the optimal rate of flow worked no matter what technique was used. However, the situation was completely different for vertical boundary cooling. A fan nozzle that sprayed the water tangent to the surface with large drops proved to be the most efficient method. Locating this nozzle at the center sideways position was more efficient than having it located either at the top or bottom. A fog nozzle producing a cone–shaped fog proved to be inadequate at the recommended rate. Performance improved with greater flow rates. However, spraying water perpendicular to the boundary surface was the least efficient way to apply the water.

Vertical cooling above the fire proved to be more difficult, and some venting was necessary for firefighters to remain there for any length of time. The 5ft (1.5m) applicator, lowered through a scuttle hole, proved to be effective for indirectly cooling the compartment above the fire compartment. Pulse tactics also worked well, with short bursts a few seconds long and with longer intervals between. The exact timing would vary with conditions.

In summary, maximum application efficiency was obtained by applying water tangentially to the boundary surface in sheets or large droplets with nozzles that produced a fan–shaped spray pattern. However, because of furniture and other materials in the compartment, it could be impossible to apply water to all of the surfaces.

In these tests, the problems in reaching the objectives centered not on equipment but on personnel. Firefighters could not stay in the adjacent compartment for long. The time ranged from as short as 5min to 10min to no longer than 20min. It is always advisable to relieve firefighters before they reach the limit of their tolerance. In addition, serious problems occurred with protective clothing, especially gloves. Firefighters were prone to receive burns around their necks, wrists, hands, ankles, and feet. Gloves were vulnerable; once wet, they had to be cooled with water frequently to prevent scalding burns.

So with proper equipment and with relief of personnel, this research did identify adequate methods for stopping the spread of a worst–case compartment fire. The methods used in this research provide an important lesson for land–based fire departments: It is not the volume of water that is crucial in stopping the spread of a fire; it is the way in which the water is applied by using fog nozzles.

Third test series

In the first two test series, no attempt was made to control or extinguish the compartment fire. This was the purpose of the third test series, conducted by Commander John P. Farley in 1994 at the Naval Research Laboratory, Mobile, Alabama, onboard the *USS Shadwell*. The objective of these tests was to develop a full–scale fire threat in a compartment that would provide conditions for comparing the safety and effectiveness of a fog attack versus a direct straight–stream attack.

The fire threat was a growing, steady state Class A fire involving wood cribs, particle–board panels, and newspaper–filled cardboard boxes. Fires were set in three different locations in the compartment so that office equipment blocked direct access from a position near the entry door. The fire was allowed to burn for 10min to 15min before entry. Upper layer temperatures reached 500°C to 600°C (932°F to 1,123°F). A 1.5in (38mm) attack line was used with a variable–pattern fog nozzle.

In the straight–stream attack, the team advanced into the compartment so they could apply water directly to all three burning areas. A short–burst tactic was used to minimize water damage and steam production.

In the fog attack, the team used the 3–D tactics developed in Europe. They entered 1.2m to 1.8m (4ft to 6ft) and took a crouching position. The nozzle was set to a medium fog pattern (60°) and discharged at a 45° angle overhead. A series of two to three short bursts, 2s to 3s in duration, was sufficient to knock down the fire. Then the team advanced to a position that enabled them to make a direct attack upon the three fire areas using a straight stream.

The third series of tests was designed to measure the safety and effectiveness of the two methods of attack in the following fire situations:

1. Growing steady–state fires where the seat of the fire is shielded from direct attack by an obstruction.

2. Growing steady–state fires where there are multiple fire sources scattered throughout the space.

3. Low–visibility steady–state fires where there are multiple fire sources scattered throughout the space.

These fires were much like what civilian firefighters encounter every day. Only one test was conducted in which there were no obstructions so that direct access to the source of the fire was available. The tests did not involve a fire that had flashed over. All of the tests were documented so that they could be repeated easily.

Five measurements were taken using the following instruments to determine the safety and effectiveness of the two methods of attack:

1. Wood crib thermocouples (measuring temperature).

2. Average of overhead thermocouples (measuring temperatures).

3. Upper and lower calorimeters (measuring heat volume).

4. Average of upper thermocouples versus average of lower thermocouples (measuring temperatures).

5. Water flowmeters (flow in gallons per minute).

These measurements established four criteria that determined the effectiveness of a given method of attack:

1. The wood crib temperatures show when the fire was knocked down and when it was extinguished.

2. The average of the overhead temperatures showed the thermal threat in the overhead and when it was controlled.

3. The measures of the total heat flux (calorimeters) and the averages of the upper and lower temperatures showed how much the thermal balance in the compartment was disturbed.

4. The flowmeter showed how much water was used for each attack. [7]

The following definitions were used to establish criteria in determining the effectiveness of the two methods of attack.

- Control time—When the attack team was able to advance from its initial attack position and begin applying water to the fire sources

- Extinguishment time—When all crib temperatures were reduced below 125°C (257°F), there was no visible flaming, and there were no subsequent reflashes reported

Conclusions

The following conclusions were made from the data produced and from the observations of those who supervised or participated in these tests.

First, there were no significant differences in total water usage between the two methods of attack. The amount used for control averaged around 10gal (38L) with a control time of < 1min. Extinguishment was achieved in only two tests with a time of 4min using < 100gal (378L) of water. The differences between the two methods of attack lay in the efficiency with which water is used rather than in the total amount used.

Second, with respect to safety, Farley concluded, "It was determined that the offensive fog attack, using a medium angle fog directed 45° upward at the flaming overhead and discharged in short bursts, appears to be an effective tactic for controlling an overhead fire threat."[9] There are several reasons for this conclusion. First, the fog attack immediately reduced the overhead temperatures by 200°C to 250°C (392°F to 482°F), and the temperatures continued to fall throughout the fire attack. The threat of flashover was eliminated, and final extinguishment was completed within 5min to 10min. None of the fog attack teams suffered any threat of burns.

In contrast, the straight–stream attack showed that overhead temperatures were reduced at first but quickly rebounded. The threat of flashover became greater, forcing the teams to retreat out of the compartment within 2min. Many team members suffered minor burns to hands, wrists, face, neck, and back.

Second, there were significant differences in the effectiveness of the two methods of attack. Because of the obstructions and the lack of familiarity with the layout of the compartment, the attack was slowed down so much that controlling the overhead fire became the top priority. The fog attack was far more effective in controlling the overhead fire. The report concludes with the following:

> When applied properly, e.g., a 60° fog stream directed at a 45° angle upward at the flaming overhead, and discharged in short, controlled bursts, 2s to 3s in duration, the offensive fog tactic can be used effectively to attack and control a fire without disturbing the thermal balance in the compartment. Generally, three bursts will be sufficient to control the overhead fire and to allow the direct attack to continue safely. The exact number of bursts to be used is dependent on the degree of control and knock down achieved. Also, throttling the nozzle to reduce water flow may provide additional benefits by reducing steam production.[10]

Third, the fog attack also provided the least disturbance to thermal balance.

Farley explained this as follows:

> The disturbance of the thermal balance within the fire compartment was best shown by comparing the total heat flux measured by the calorimeters mounted 0.9m and 2.4m (3ft and 8ft) above the deck in the fire compartment. The key indicator of significant disturbance in the thermal balance was the upward spike in the 0.9m (3ft) heat flux that approached or met the heat flux for the 2.4m (8ft) calorimeter. This indicated total compartment mixing with steam.[11]

Figure 3–2, showing FOG–12 (straight stream attack), depicts thermal imbalance. Notice the vast difference between the straight–stream attack and the fog attack (FOG 15). The fog attack reduced the temperatures at lower and higher levels within 2min, followed by a steady decline for the next 10min. By contrast, the straight–stream attack did not bring the temperatures under control until about 12min. Also notice the spikes in the temperature at the 3ft (0.9m) level that approach the temperature at the 8ft (2.4m) level. This indicates extreme turbulence in the compartment.

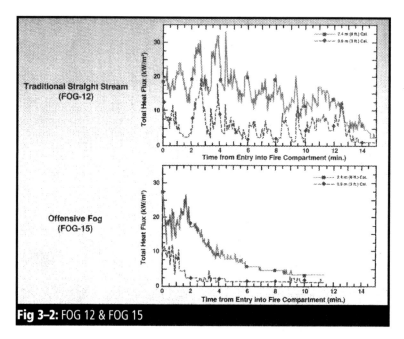

Fig 3–2: FOG 12 & FOG 15

All of this is dramatic confirmation of the research done at Iowa State University by Bill Nelson and Keith Royer. A fog nozzle used properly can be used safely and effectively to maintain thermal balance. Both of these research efforts reached this same conclusion.

The traditional straight–stream attack was not nearly as effective as the fog attack. Deflecting the straight stream off the bulkhead and overhead created excessive amounts of steam. If attack teams had to back out of the compartment, it was because of the threat of steam rather than heat. The fog attacks resulted in no burns to team members, whereas the straight–stream attack resulted in burns to the hands, wrists, face, neck, and back. The report concludes the following:

> Tenability (in terms of heat stress to firefighters) in the space was improved when the offensive fog tactic was used. Using a straight stream or narrow angle fog directed to the overhead resulted in excessive amounts of hot, penetrating steam. The steam generated by the offensive fog was not as severe and resulted in no burns to firefighters.[12]

Summary

In the 1990s the U.S. Navy conducted research in six phases. In addition to the three phases considered here, the Navy also investigated the impact of Navy fire insulation, the venting of large shipboard fires, and fire dynamics. A massive amount of data was accumulated in all of these experiments.

In one of the tests in the third series, the obstructions were removed so that a direct attack upon the fire was possible from the entry door. This test showed no significant advantage for either the offensive fog attack or the traditional straight–stream attack. Both were equally effective. This is a highly significant finding that should end the controversy raging today in the fire service. For this particular type of fire (pre flashover two–layer confined fire), neither method of attack was any better than the other.

Finally, we should note that the research done by the Naval Research Laboratory and at Iowa State University converge on the same answers on how to use fog nozzles safely and effectively. If you use the right amount of water with the 3–D pulse tactic or a combination attack, either method works effectively. Once again, all of these findings illustrate a fundamental principle of fire combat: No one method of attack can possibly solve all your firefighting problems. Each method is useful and is suitable for one type of fire or another. Truly there is no magic pill for the fire service.

The three test series considered here established a number of significant findings. This research produced a design fire with the maximum rate of heat release for a compartment fire. It also produced two conditions needed for flashover to occur. The optimum rate of flow for stopping the spread of fire was identified. The research also identified the need for rapid rotation of team members to avoid heat stress. The weaknesses of protective clothing were identified—especially the poor performance of gloves. And the tests proved the superiority of a fog attack in cooling overhead gases where a direct attack upon a compartment fire is not possible from the doorway. This confirms the research done in Europe with respect to the 3–D gas cooling pulse tactics.

Notes

1J.T. Leonard et al., *Post–Flashover Fires in Simulated Shipboard Compartments, Phase Small Scale Studies* (Naval Research Laboratories, 1991), p. 2.

2Ibid, p. 4.

3Ibid, p. 4.

4Ibid, p. 51.

5R.L. Darwin et al., *Post–Flashover Fires in Shipboard Compartments Aboard Ex–USS Shadwell, Phase VI – Boundary and Compartment Cooling* (Naval Research Laboratory, 1994), p. 1.

6John P. Farley et al., Phase II – *Full Scale Offensive Fog Attack Tests* (Naval Research Laboratory, 1997), p. 3.

7Ibid, p. 39.

8Ibid, p. 90.

9Ibid, p. 78.

10Ibid, p. 96.

11Ibid, p. 79.

12Ibid, p. 96.

CHAPTER 4

European Research
(Courtesy of Paul Grimwood)

Something very important has occurred. The conclusions being drawn from research in every country—the United States, England, Finland, Sweden, New Zealand, Australia, and others—are virtually identical. All of this research has reached the same general conclusion: fog nozzles are the safest and most effective tool for certain types of fires.

For example, when the mathematics were translated from the metric system, the conclusions drawn from the research at Iowa State University and the US Naval Research Laboratory were no different from the Swedish and the Finnish results. For the first time in more than 250 years of organized firefighting, knowledge is being standardized. This is not surprising. During the past 15 years, researchers have come to possess both improved research methods and better instrumentation. The mathematics used to describe and control this research has also been refined to better serve the profession.

The organization of European fire departments (brigades) differs widely from US counterparts, both in their tactical deployment and fire ground operations. Cross training and staffing practices are standardized throughout Europe to a greater extent, with all engines providing a complement of four to six firefighters and aerial trucks crewed by two to three firefighters, with additional firefighters remaining unassigned to specific company roles until arrival on the scene, similar to FDNY squads. While appreciating the benefits of venting actions during structural firefighting, it is not common to see such a strategy employed at an early stage of the fire, and greater emphasis is placed upon a prompt siting of attack hose lines and controlling interior conditions by reducing, where possible, air tracks (gravity currents) entering the structure and feeding the fire.

The water supply systems provided throughout Europe for firefighting purposes are generally older, smaller diameter, and less capable than those in the United States. The hydrant grids are operated at lower pressures through narrow channels, and the fire engine pumps are less technical, while hoses used to transport water onto the fire are generally of lower strength and capacity. However, Europeans have adapted well, and they commonly utilize low-flow, high-pressure 500psi (34bar) hose-reel (booster) lines of 19mm or 25mm (0.75in or 1.0in) to tackle 80% of their structure fires, using less than 370gal (1,400L) of water carried on each engine. It is common to see two of the high-pressure lines laid in before resorting to higher flow 1.75in (44mm) or 2.75in (70mm) low-pressure lines. Much emphasis is placed on a speedy attack and the quick-water concept of using tank water through lightweight, low-flow 30gpm (113.5Lpm) high-pressure lines. This is well suited to the multistory brick and concrete construction common throughout Europe.

The tactical approach to structural firefighting based upon the concepts of Layman's indirect attack has evolved over many years, although direct extinguishing using narrow fog patterns and straight streams are equally as common. This attack strategy is very aggressive; streams are applied from the interior, working close to the fire. This has caused firefighters to suffer from steam envelopes and superheated environments in the past; nevertheless, the fires have generally been extinguished.

In 1982, following the loss of two firefighters' lives in a Stockholm flashover, the Swedish Fire Service introduced an innovative adaptation[1] of the Layman principles[2] while working from an interior position. It negated all of the problems previously associated with close-quarters fire combat and the compartmental use of water fog. The new-wave applications were termed *offensive firefighting* and, later, *3-D water fog attack* when it was realized that the objective was to apply fine water droplets directly into the fire gases overhead. This 3-D effect of applying water fog overhead was, in real terms, different from the Layman approach, which intended evaporation to occur as water came into contact with the hot surface, walls, ceiling linings, etc., while targeting a 10% mix of water vapor within the compartment. Unlike the Layman approach, the effects of 3-D water fog applications relied far less on the smothering effects of excessive water vapor and more on the cooling effects that occurred in the gases.

Anders Lauren originated the idea in 1982 in Stockholm, Sweden, of using a steel shipping container lined with panels of chipboard to demonstrate how fire gases formed inside a fire compartment before they eventually ignited as a flashover simulator to train firefighters in the Stockholm Fire Brigade. This innovative use of a simple, cheap, and freely available structure was further adapted for safety and effect and became known internationally as the *flashover simulator*, or *can*. The design of the simulator has been researched, developed, and used worldwide to teach fire behavior to firefighters while enabling them to practice various nozzle techniques to deal with the fire gases building toward the rollover and flashover stages of development.[3]

In 1984, the use of 3-D water fog applications were researched operationally by firefighters in London with much success,[4] although it was some years later before the strategy became officially accepted by the UK Home Office. Unfortunately, it wasn't until three UK firefighters were killed in two separate flashovers within a few days of each other in 1996 that the new-wave strategy was finally approved in 1997.

By the turn of the new millennium, the new-wave uses of 3-D water fog were being researched and deployed by many fire authorities across Europe, Australia, New Zealand, and South Africa. The new-wave applications developed throughout Europe since the 1980s were particularly effective when traversing the approach route to fire-involved compartments. The improvements in visibility and maintenance of thermal balance when compared to smooth-bore attack were particularly outstanding. On a few occasions the fire would grow so rapidly that no amount of 3-D fog application would control it. Even a smooth-bore application failed to reduce the fires' intensity, causing firefighters to retreat. This would suggest conditions where air tracks remained uncontrolled and heat release rate (HRR) overpowered the lines and flows in use.[4]

The loss of several firefighters' lives in flashover-related events prompted the need for a review of firefighting strategy in Europe. Firefighters did not have a clear appreciation of how compartment fires developed under varying ventilation parameters, and little attention was being directed to the dangerous formation of flammable and explosive fire gases overhead. Sometimes these gases would transport into adjacent compartment or voids, some distance from the fire, and ignite under delayed circumstances—occasionally after the main fire had been suppressed. The current definition of flashover, backdraft, and smoke explosion were closely researched by scientists in the UK,[5] Sweden,[6] New Zealand,[7] and the United States,[8] and several events associated with rapid fire progress were clarified. This knowledge encouraged a renewed approach to compartment firefighting, and the flashover simulators (containers) were used to teach fire behavior, fire compartment entry techniques from adjacent areas, and 3-D pulsing actions at the nozzle to control conditions overhead.

3-D Fog Attack

An application of water fog discharged in short, controlled bursts (pulses) where the water droplet range is critical. There are three main uses of 3-D water fog in compartment firefighting:

1. To cool the gases overhead below temperatures that support any progression to rollover or flashover conditions.

2. To create an inert atmosphere overhead—one of controlled steam or suspended water droplets where no evaporation occurs—to prevent or suppress/mitigate any likely ignition of these gases in a backdraft or smoke explosion.

3. To suppress burning reservoirs of fire gases that have accumulated and are burning off in geometrical voids such as stair shafts, attics, or compartments.

The objective of the first and second uses (defensive) is to suspend the droplets into the fire gas layers to cool, inert, and dilute them, bringing them outside their immediate range of flammability to prevent or quench subsequent ignitions. The objective of the third use (offensive) is to knock down fire-gas formations that are burning off inside a compartment under a ventilation-controlled regime. The application is made on a 3-D basis into a cubic volume of fire gases inside an enclosure, and the strategy is applied at close quarters, with firefighters occupying the compartment at the time of attack. This application of water fog may be used inside both vented and unvented compartments.

These applications are used not solely to extinguish fires but mainly to make safe the approach route to the fire and reduce the likelihood of fire gas ignitions. Neither are these techniques designed to replace the direct style of fire attack utilizing water in a straight-stream setting; rather, they complement existing forms of fire attack to increase the safety and effectiveness of fire-fighting teams.

An interesting research project undertaken by Lund University, Sweden, used a computer model to demonstrate the likely effects of actions taken at an underventilated compartment fire using natural ventilation, PPV, and 3-D water fog applications.[9] The study's conclusions demonstrated the 3-D water fog application as the safest option of the three.

Three-dimensional tactics may be either defensive or offensive. The true qualities of 3-D water fog applications are realized in flashover and backdraft prevention (defensive). Pulsing water fog overhead on the approach route, using short, rapid bursts at the nozzle, serves to inert the fire gas layers and prevents or mitigates the potential for any ignition of the fire gases that may lead to such a major event. Such ignition of accumulated fire may vary in their explosive force, but fine water droplets offer a quenching effect under such conditions and lessen the explosive effects. Again, the applications are administered with a reasonable amount of precision and are dependent on equipment, firefighter awareness, and training. The duration of pulses and the degree of cone spread varies according to the size of the compartment and the conditions present.

Contraction of gases

There is a fundamental reason why the 3-D fog attack is safe and effective in controlling a dangerous overhead full of fire gases. If the pulse tactics are used, then the overhead gases will be contracted compared to the previous volume before the fog attack. This contraction is extremely important. It means that firefighter safety is preserved—they will not be engulfed in steam. It also means that thermal balance is preserved so the firefighters can proceed to extinguish the fire by a direct attack. To achieve this concentration, the nozzle operator must use the proper fog cone and application angle as well as pulse the nozzle correctly.

Given these conditions, the following example is proof that this contraction actually does occur.[10] Suppose that a 60° fog cone is applied at a 45° angle to the floor into a 50m³ (1,766ft³) room. A 1s spurt from a 100Lpm (26gpm) flow hose line will place approximately 1.6L (0.4gal) of water into the cone. For this explanation, let's use a single unit of air heated at 538°C (1,000°F), weighing 0.45kg (1lb) and occupying a volume of 1m³. This single unit of air is capable of evaporating 0.1kg (0.1L or 0.22lb) of water, which, as steam (generated at this, a typical fire temperature in a compartment bordering on flashover), will occupy 0.37m³ (13ft³). A 60° fog cone, when applied, occupies the space of 16 units of air at 538°C (1,000°F). This means that 1.6kg (16 x 0.1kg, or 1.6L; 3.5lb) of water can be evaporated (i.e., the exact amount discharged into the cone during a single one second burst). This amount is evaporated in the gases before it reaches the walls and ceiling, maximizing the cooling effect in the overhead. Where droplets are oversized or over drenching occurs, too much water will pass through the gases to evaporate into undesirable amounts of steam as it reaches the hot surfaces within the compartment.

By using calculations involving volume, pressure and temperature of gases, we are able to observe how the gases have been effectively cooled, causing them to contract. Each unit of air within the cone has now been cooled to about 100°C (212°F) and occupies a volume of only 0.45m³ (16ft³). This reduces total air volume (within the confines of the cone's space) from 16m³ to 7.2m³ (565ft³ to 254ft³). However, to this we must add the 5.92m³ (209ft³) of water vapor (16 x 0.37) as generated at 538°C (1,000°F) within the gases. The dramatic effect creates a negative pressure within the compartment by reducing overall volume from 50m³ to 47.1m³

(1,766ft³ to 1,664ft³) with a single burst of fog. Any air inflow that may have taken place at the nozzle will be minimal (about 0.9m³, or 31.8ft³), and the negative pressure will be maintained. Overall, there is no expansion whatsoever; however, the mass of gases are not stable and are constantly in motion and in a state of transition. It is important, therefore, that nozzle operators continually assess conditions following each burst, or series of bursts (pulses), so adjustments in pulse duration and cone pattern may be made.

This is proof that if you use the right amount of water, then there is no "pushing of a fire" anywhere. On the contrary, there is a contraction that preserves thermal balance. This enables the firefighters to continue the fire attack with safety. This is absolute vindication of Bill Nelson's statement of the fundamental principle of firefighting: "All one needs to do is put the right amount of water in the right place, and the fire is controlled". How true this statement is.

The annual BFRL Conference on Fire Research in 1998 produced an interesting (NISI) paper from Alageel, Ewan, and Swithenbank, University of Sheffield, UK, who investigated the "Mitigation of Compartment Jet Fires Using Water Sprays". The main objective of the study was to investigate the interaction of water sprays with a ceiling jet fire in a ventilation-controlled state. Close attention was paid to the effectiveness of different spray angles, droplet diameters, stream velocities, and water flow rates. It was generally observed that water applications into the gas layers utilizing different spray angles of 30°, 60°, 75°, 909°, 120°, and 150° produced varying reductions in compartmental temperatures. But spray cones within the 60° to 75° range were the most effective in reducing the overall temperature.

For these angles the limiting behavior due to the effectiveness in penetrating the flame indicated that spray velocities in excess of 18m/s (40mph) should be used. The mean droplet diameters of 100µ to 600µ (0.1mm to 0.6mm or 0.004in to 0.024in) were analyzed. It was further noted that droplets within the 300µ (0.3mm or 0.012in) range maximized any cooling effects within the compartment. In terms of flow rate, it was reported that for these compartmental dimensions (which were the same as a standard simulator container of 35m³ (1,237ft³), the optimum flow rate was 120Lpm to 180Lpm (32gpm to 48gpm). Where this flow rate was exceeded, the compartmental temperatures were not reduced any quicker and much water was observed as runoff. At flow rates below 120Lpm (32gpm), the overall cooling of the gases was much less effective.

As an extinguishing medium, water had a theoretical cooling capability of 2.6MW/L/s (2,464Btu/s).[11] It is prudent to try to match flows with the likely heat release rates that may be encountered on initial entry in local structures. The average one-room residential fire is likely to reach intensities exceeding 7MW at flashover, and a minimum flow of 500Lpm (132gpm) will be required to handle this situation safely and effectively. However, such a flow rate is too high for an optimized gas-cooling application. A flow of around 100Lpm to 150Lpm (26gpm to 40gpm) will be more suited to the same fire during its pre flashover stage where gas cooling/inerting is relevant. To avoid bringing in larger streams and playing catch-up as the fire escalates, the firefighter might ideally be equipped with an initial attack hose line that provides a flow range of 100Lpm to 500Lpm (26gpm to 132gpm) with a selectable flow option at the nozzle. An alternative combination nozzle of fixed flow, or automatic type, may be used where a flow control facility allows pulsing actions at lower flows just by cracking the flow handle/trigger on and off.

Using a computer model, Rasbash attempted to estimate the heat transfer between flames and water sprays and produced a plot of convective heat transfer rate against drop velocity for drop sizes ranging from 50µ to 2mm (0.002in to 0.08in) while assuming a flame temperature of 1,000°C (1,832°F). In general, higher velocities and smaller droplet diameters increased the heat transfer rates. For example, a 2mm (0.08in) drop at 0.07m/s (0.23ft/s), terminal velocity is still air, produced a heat transfer rate of 167kW/m² (14.7Btu/ft²/s). The same drop traveling at 2m/s (0.65ft/s) achieved a value of 293kW/m² (25.8Btu/ft²/s). For a 50µ (0.002in) drop at velocities of 0.01m/s to 0.5m/s (0.3ft/s and 1.6ft/s), the corresponding heat transfer rates were 1.7MW/m² and 2.5MW/m² (150Btu/s and 220Btu/s), respectively. The research also noted that drops of larger initial size were able to penetrate farther into the flame before complete evaporation occurred.

Benefits summary

The following statements summarize the benefits of using the 3-D water fog pulse tactics.

1. 3-D water fog tactics have been proven scientifically to be the most effective way to cool gases overhead in comparison to any other form of fire attack; including smooth-bore, indirect fog, Class A foam, and CAFS methods. This is supported by several independent studies around the world.

2. A 3-D water fog, correctly applied, will have an inerting effect on the gases, rendering them less likely to ignite by reducing the oxygen partial pressure by adding an inert gas (*e.g.*, N_2, CO_2, or H_2O vapor). This is li. e removing the oxidizer supply to the flame by producing water vapor. This is the dominant mechanism by which water mists can suppress large, confined fires.

3. The injection of water droplets into fire gases narrows their limits of flammability and further reduces the likelihood of ignition.

4. Cooling the flame zone directly reduces the concentration of free radicals—in particular, the chain–branching initiators of the combustion reaction. Some proportion of the heat of reaction is taken up by heating an inert substance, such as water; therefore, less thermal energy is available to continue the chemical breakup of compounds in the vicinity of the reaction zone. One function of the new water mist technology is to act in this manner, with the fire droplets providing a very large surface area per unit mass of spray to increase the rate of heat transfer.

Myths and misconceptions

There are many ill–informed arguments and misconceptions mounted against the tactical use of 3–D water fog developed in Europe.

The stream from a smooth–bore nozzle can be used just as effectively to cool gases overhead by utilizing a Z–pattern. This is a myth. It has been scientifically proven in several independent research studies that fine water droplets will cool gases overhead far more effectively than a straight–stream application. The US Naval Research Laboratory research, besides the European research, clearly demonstrated this fact under strict scientific monitoring.

The application of water fog causes steam burns to firefighters and pushes fire ahead of the stream. This will not happen where a pulsing action is used at the nozzle, using short bursts to place about a cupful of water droplets overhead with each brief pulse. The water will then evaporate in the gases and not on superheated surfaces such as walls and ceiling. This cooling effect causes the gases to contract, and steam production is dry as opposed to cloudy wet. There is not enough force from the pulses to push fire ahead of the stream.

The use of water fog upsets the thermal balance. The actual effect where 3–D applications are used is exactly the opposite. The smoke layering is maintained and visibility is optimized by pulsing water fog overhead. This has constantly been demonstrated in scientific studies, including the US Naval Research Laboratory tests that compared the stream from a smooth–bore nozzle.

The flow rate required for gas–phase cooling is dangerously low. But with a select–a–flow combination nozzle, you have higher flows immediately available at the flick of a switch. Gas cooling can also be affected with high–flow automatic nozzles. You must adjust your applications to suit the equipment and flow rates you normally work with.

Water fog tactics are optimized in unventilated spaces. This is certainly not true of 3–D water fog applications, which are effective in both ventilated and unventilated compartments.

The pulsing action at a nozzle may create dangerous water–hammer effects. The views of major fire pump manufacturers suggest there is no danger of causing damage to fire pumps by using pulsing actions at the nozzle. Since introducing these techniques in Sweden, the UK, and Australia, there has been no noticeable increase in pump/nozzle maintenance and repairs, although there have been some problems with bursting hose reels (booster lines) when pulsed at 500psi (34bar) pump pressures. This problem has been resolved. There are also engineering solutions available in terms of pressure relief valves and hydraulic retarders that can be fitted to pumps and nozzles where any concern exists.

The tactical solutions and training implications associated with applying water to control environmental conditions within a fire compartment/structure go far beyond nozzle techniques. The training concepts create a greater awareness of fire growth and development, fire behavior patterns, formation and behavior of flammable fire gas layers, environmental and tactical risk assessment, the effects of compartmental geometry and layout, and the tactical approach to varying situations, including opening and entry procedures and the effects of tactical venting actions. The style of approach is being adopted worldwide, with firefighter safety cited as the prime concern.

This research has produced highly significant results. Once again, these results converge upon the same set of scientific facts and principles. The ideal water droplet diameter for gas–phase cooling is identified as from 200μ to 400μ. The spray–cone size most effective in reducing temperatures was 60° to 75°. The ideal flow rate was between 120Lpm and 180Lpm (32gpm and 48gpm).

Where this flow rate was exceeded the compartmental temperatures were not reduced any quicker and much water was observed as "run–off" whereas at flow rates below 120Lpm (32gpm) the overall cooling of the gases was seen to be much less effective.

Once again, this confirms the fundamental principle cited by Bill Nelson of Iowa State University: that all one needs to do is put the right amount of water in the right place and in the right way the fire is controlled.

Scandinavian Research

In 1995 a four–year research project was completed by Finland's Fire Technology Laboratory (VTT).[12] Dr. Maarit Tuomisaari used a computer analysis and live fire tests to study the fire–suppressive qualities of water sprays when applied into gaseous combustion in compartmental firefighting. The research compared indirect applications onto hot surface linings using sweeping motions against intermittent 3–D bursts (pulses) directed into the burning gases of post flashover fires. The amount of water used and the average water droplet size were the two most influential factors to affect fire control times. In line with many other studies of this nature, the droplet range of 200μ to 600μ (0.2mm to 0.6mm or 0.008in to 0.024in) was the most effective way to suppress the burning gas layers. While indirect sweeps of the linings effectively cooled and suppressed the burning gases, the disruption of the thermal layer was an undesirable effect when compared to intermittent pulses applied directly into the gases, where thermal disruption is nonexistent and a positive balance is maintained. The use of intermittent pulses of water fog also optimized the actual amounts of water injected overhead, helping the nozzle operator maintain control of conditions and reduce undesirable steam expansion.

Water being sprayed into the fire compartment can generally be divided into three main parts. One part of the water (small droplets) is blown away through failure to penetrate the updraft in the compartment and thus does not participate in the suppression. A second part is vaporized (ideal droplets) in the combustion gases. And a third part (large droplets) reaches internal surfaces and the fuel in liquid form, where it is vaporized or flows to pool on the floor. To optimize 3–D gaseous fire suppression, the amount of vaporization must be maximized. To ensure the vaporization effects are positive toward firefighters and victims occupying the compartment, most vaporization should occur in the gases and not on wall or ceiling linings. The resulting contraction of the gases will overcome any expansion of the water vapor, providing droplets within an acceptable range so that the nozzle operator is not overzealous in the application of water.

In 2000, a further research project was completed in Sweden at the request of the Stockholm Fire Brigade[13]. Anders Handell of Lund University evaluated various firefighting fog/spray streams using computer–aided technology and live–fire experience to compare the effectiveness of a wide range of nozzles in cooling the superheated gaseous conditions that exist overhead in a fire–involved compartment. This was also the objective, in part, of the earlier VTT research. Both research projects concluded that the most effective nozzle pattern for gas cooling and burning gas suppression was provided by equipment from Task Force Tips (USA)[14]. As a result of this research, the Stockholm Fire Brigade initiated a nozzle–replacement program in 2001 to change to the TFT Ultimatic, a nozzle also used by the London Fire Brigade since 1992. The Lund research again paid close attention to water droplet size, stream patterns, flow, and velocity of firefighting nozzles as well as application techniques. This was a turning point, in that the North American nozzle outperformed the TA Fogfighter, which had been considered as the most effective nozzle for 3–D gas–cooling applications throughout the Swedish fire service.

Beyond a doubt, the transition to 3–D offensive–style water fog attacks using pulsing applications of fine water droplets into the superheated and gaseous overhead has saved firefighters' lives. Statistics have demonstrated that new–wave methods of preventing or reducing the potential for igniting forming gas layers in subsequent rollovers, flashover, backdrafts, or smoke explosions while dealing most effectively with post flashover burning gas reservoirs have drastically reduced the death and injury rates of firefighters caused by such rapid fire propagation.

Remember, this new–wave use of water fog in compartmental or structural firefighting complements traditional fighting methods, such as direct straight–stream attack. The firefighter who is able to assess the risk and recognize the application that is optimal for the conditions as they present themselves is the one most likely to succeed.

Flow Rates

Several international research projects over the past 50 years have attempted to produce an engineered solution to water flow–rate requirements for structural firefighting. These studies generally have been based upon scientific data associated with heat release rates from compartment fires, along with empirical research investigating actual flow rates used by fire brigades when tackling fires in a wide range of occupancy types. This information is most useful for grading firefighting water flow requirements in line with building codes. It is also of use to the operational fire officer who must assess the resources required at a particular incident to suppress any structure fire of a known estimated size.

Before Grimwood's research study in 1990,[15] the most established research to date had been completed in the United States, although there had been several small–scale laboratory studies investigating theoretical flow rates to suppress minor compartment fires. The conclusion of Grimwood's research, covering 100 major fires in London from 1989 to 1990, demonstrated a recommended flow rate that appeared controversially low in comparison with those currently used in the United States and caused a debate that prompted further research. This research occurred between 1994 and 1997 when Lund University supported the London Fire Brigade in a 307–fire study that culminated in the Sardqvist report (7003 report) in 1998.[16] The flow rates reportedly used by London firefighters in this study were substantially higher than those Grimwood had calculated in 1990. But why did this happen? Were Grimwood's findings somehow underestimated, or did the Lund 7003 report produce an overestimate?

The nozzle flow estimates provided by the London Fire Brigade, upon which. Sundqvist based his calculations, were not representative of actual flows achieved on the fire ground. In fact, these theoretical and unrealistic nozzle flows actually resulted in the Lund 7003 flow–rate curve being 40% too high. As a serving operational firefighter in London during part of this research period, Grimwood is able to attest that the flow rate detailed in SRDB codes at that time were rarely, if ever, achieved on the fire ground due to a number of factors, including hydrant flow capabilities, friction losses, and nozzle reaction forces. There is no mention in the codes of nozzle/hose sizes or reaction forces that would have had direct impact on the amount of water an interior attack hose line could effectively flow. For example, any attempt to flow an attack hand line at 870Lpm (230gpm) would produce a nozzle reaction force that could not possibly be handled safely by an interior attack team.[17] Further, to suggest that pressures of 5bar (72.6psi) are regularly achieved at the nozzle is generally unrealistic and often impractical because UK firefighters traditionally under pump their attack hose lines with pump pressures of 4bar to 5bar (58psi to 72.6psi). Grimwood's experience at that time would suggest that maximum flows of 200Lpm (53gpm) from a 12.5mm (0.5in) nozzle, 450Lpm (119gpm) from a 20mm (0.79in) nozzle, and 700Lpm (185gpm) from a 25mm (1in) nozzle on interior attack hose lines were far more realistic than those suggested by the SRDB codes used in the Lund research. At large incidents the flow rates may even have fallen below these estimates due to hydrant capability at the grid.

In 1994 a further study completed by Barnett in New Zealand produced scientific data, (18 supported by much empirical research), that provided a foundation for the MacBar Fire Design Code in 1997. This research produced a flow graph that is closely correlated to Grimwood's earlier work. Interestingly, the Lund 7003 flow graph is amended to demonstrate a 40% overestimate. This, too, falls much more in line with both the Barnett and Grimwood research findings.

When converted to an area formula, Grimwoods's original calculation, based on a mean average for comparison to Lund 7003, for minimum (and realistic) fire ground flow–rate requirements, based on office compartments with 2.5m (8ft) high ceilings, suggests that $A \times 2$ = liters per minute, where A = area in square meters of floor space. Interestingly, the Barnett 1994 and Grimwood 1990 studies demonstrated a flow curve directly proportional to the area of the fire and not roughly proportional to the square root of the area of the fire, as suggested by Sardqvist in 1998.

Notes

[1]Mats Rosander and Krister Gielson, *Fire Magazine UK* (1984), pp. 43 – 46.

[2]Lloyd Layman, *Attacking and Extinguishing Interior Fires* (Quincy, MA: NFPA), 1955.

[3]Paul Grimwood, *Fog Attack* (Redhill, UK: DMG World Media Publications).

[4]Ibid.

[5]Grant & Drysdale, FRDG, 1/97, UK.

[6]Richard Chitty, FRDG 5/94, UK.

[7]Lars–Goran Bengtsson, Lund University Report 12019, 1999.

[8]B.J. Sutherland, University of Canterbury, Report 99/15, 1999.

[9]Charles Fleischman, University of California, Berkley, NIST–GCR–94–846, 1994.

[10]Danaiel Gojkovic and Lasse Bengtsson, Lund University, 2001.

[11]Paul Grimwood, op. cit.

[12]Maarit Tuomisaari, *Suppression of Compartment Fires with a Small Amount of Water,* VTT, Finland.

[13]Anders Handell, Lund University Report 5065, (2000).

[14]TFT Ultimatic Nozzle.

[15]Paul Grimwood, op. cit.

[16]Stefan Sardqvist, Lund University Report 7003, (1998).

[17]Fire Magazine, (1992), p. 16.

[18]C.R. Barnett, MacBar Fire Design Code (Auckland, New Zealand).

CHAPTER 5

The Scientific Foundation For Fighting Fires

The fundamental principle in all of science is that matter can be neither created nor destroyed. Matter can only be transformed into another state or transported to another place. That a fire destroys a house is true in one sense only. The materials that formed the house no longer exist as such, but these materials have not been destroyed. They still exist in some other form or in some other place, largely as gases or ashes. Of course, this is little comfort to the homeowner.

The same principle applies to energy. Energy can neither be created nor destroyed. Thus, energy can only be transformed into other forms of energy. Using this fundamental principle, we will explain the combustion process qualitatively. That is, we will not get involved with mathematics or engineering, all of which involves a quantitative perspective.

Heat Energy

Heat is a form of energy—kinetic energy produced by particles in motion—and is classified as thermal energy. There are two aspects of heat energy: temperature and quantity (volume). A scale in degrees measures temperature, with a higher temperature indicating a greater degree of kinetic energy. Quantity of heat is measured by the amount of heat that produces a rise of 1° in a given unit of mass.

The concept of heat can be deceptive. In the 19th century, scientists believed that heat was a substance, called caloric. This substance was a massless, odorless, colorless fluid that flowed from one substance to another, thereby heating it. Scientists now know better. Heat is not a substance; it is a process by which the internal energy of a substance changes. One such change occurs when a hotter body comes in contact with a cooler body. The temperature of the hotter body always decreases while the temperature of the cooler body always increases. This process is called *homeostasis*. All adjacent temperatures will equalize if not insulated from one another.

Another change occurs as a substance undergoes a phase change, such as from a solid to a liquid or to a gas. If, for instance, the temperature of ice reaches 32°F (0°C) and enough heat energy is added to melt the ice to change it to liquid, this additional heat will not result in an increase in the temperature of the liquid water above 32°F (0°C). When we are dealing with another water phase change from liquid water to steam (gas), the term then commonly used to refer to this is the latent heat of the vaporization of water. This is the amount of heat needed to form steam and that must be retained to remain steam. If this heat is lost to the surrounding area, then the steam immediately condenses and forms liquid water. However, this common term is not accurate because it implies that there are two kinds of heat:

1. Sensible heat that we can feel

2. Latent (hidden) heat that we cannot feel

Most certainly, there are not two kinds of heat. In fact, there is not even one kind of heat. Heat is not a substance. So the preferred scientific term to use is the 'enthalpy of vaporization of water'. No attempt will be made to define "enthalpy", but it relates to the internal energy of a given substance.

Again, the additional energy needed to form steam does not result in an increase of the temperature of the steam above 212°F (100°C). This scientific fact is critical in understanding how water is used to fight fires.

Combustion

Combustion is a rapid chemical combination of a substance (fuel) with oxygen that usually produces both heat and light. The fires that we fight are mostly a hydrocarbon – air diffusion flame process. *Hydrocarbon* identifies the fuels as substances that contain hydrogen and carbon. Air identifies the source of oxygen, 21% of which is in air. *Diffusion flame* is the process whereby fuel gases mix with oxygen in the air in a complex series of chemical reactions. The flames are where heat and light are emitted.

The first theory of fire is usually presented as a fire triangle—a closed figure with three sides (see Fig. 5–1). This indicates the three elements that must be present for a fire to occur.

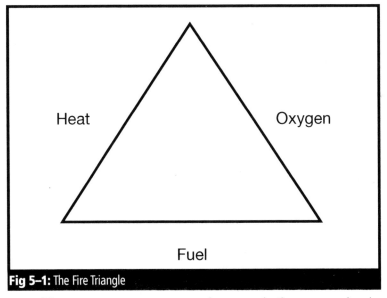

Fig 5–1: The Fire Triangle

The most common source of oxygen is the oxygen in air. Sometimes there may be another source of oxygen (an oxidizer) for a fire. In any case the oxidizer must be in the proper form, that is, present as a gas or vapor. Further, the oxidizer must be present in the proper amount. This means that the fuel–to–air mixture must be within the flammability limits for the given fuel for combustion to occur. The fuel must be present in the proper form and in the proper amount and, like the oxidizer, must be present as a vapor or gas. If the fuel–to–air mixture is above the upper flammability limit, combustion will not occur and the mixture will be too rich (too much fuel). If the mixture is below the lower flammability limit, combustion will not occur and the mixture will be too lean (not enough oxygen). Within the flammable range, combustion is ready to begin, needing only the energy necessary for ignition. Essential to the combustion process is the fact that fuels must be heated and vaporized before fire begins.

We live in a world in which fuels and oxidizers are present everywhere, so fire prevention is mainly concerned with preventing sources of energy from becoming sources of ignition. On the other hand, if a fire has occurred, fire can be suppressed by removing any component of the fire triangle.

However, there is one weakness in this theory. The fire triangle cannot explain how the combustion of organic fuels can be extinguished by using dry chemicals. Dry chemicals do not remove the energy side of the triangle, the oxidizer side, or the fuel side. So the triangle theory breaks down, and a new theory must be formulated to explain how fires can be extinguished in all cases. That theory is the fire tetrahedron (see Fig. 5–2).

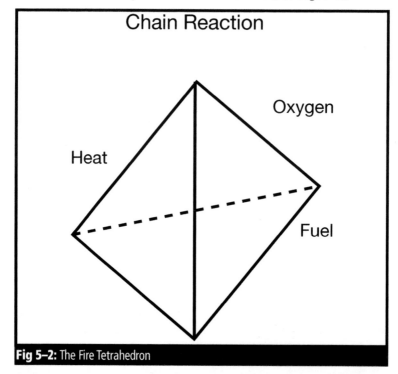

Fig 5–2: The Fire Tetrahedron

A tetrahedron is a 3–D figure with four triangular faces, or sides. The same relationship exists among these four sides as for the fire triangle. All four components must be present at the same time and place and in the proper form and amount for fire to occur. Three sides of the tetrahedron are the same as for the fire triangle. The fourth side is the chain reaction of burning, or the formation of free radicals in the fire. However, there is a weakness in the tetrahedron theory. Free radicals are not formed until after the combustion process begins. Nevertheless, the chain reaction that forms free radicals is extremely fast, with a time interval of milliseconds.

What are free radicals and how are they formed? All matter is composed of three substances:

1. Pure substances, called elements

2. Compounds formed by chemical reaction of two or more elements

3. Mixtures of elements and compounds

There are 103 elements in nature. All elements are made up of atoms, which are the smallest particles of an element that are unique to it.

Atoms are composed of three kinds of particles. The first are protons in the nucleus, each with an electrical charge. The second are neutrons, also located in the nucleus but with no electrical charge. The third are electrons, each having a single negative electrical charge that is located in orbit around the nucleus. An atom has the same number of positive protons and the same number of negative electrons; hence, it is electrically neutral. Each element has a different number of protons and neutrons that gives it a unique chemical identity. The atom is very stable, even though it may gain or lose a few of its electrons and share some of its electrons with other atoms. The nucleus is also very stable and is unaffected by fire.

One important characteristic of an atom is its atomic weight. The atomic weight is equal to the number of protons and neutrons in the nucleus. Hydrogen has the smallest atomic weight of one, with only one proton and one electron. The heaviest natural element is uranium, with 92 protons, 146 neutrons, and an atomic weight of 238. An isotope of U–238, U–235, has only 143 neutrons.

Elements are classified as metals or nonmetals. There are 81 metallic elements and 22 nonmetallic elements. Four non-metallic elements play a key role in combustion, as shown in Table 5–1.

Table 5–1: Elements That Play A Role in Combustion and Their Atomic Weights	
Element	**Atomic Weight**
Hydrogen (H)	1
Carbon (C)	12
Nitrogen (N)	14
Oxygen (O)	16

Hydrogen, nitrogen, and oxygen are diatomic, that is, they exist in nature as two atoms chemically combined to form molecules: H_2, N_2, and O_2.

Another grand division of elements and compounds is the classification as either organic or inorganic compounds. Organic compounds contain carbon. Inorganic compounds consist of minerals and salts and the compound formed from them.

There is a crucial difference between the chemical bonds formed by organic and inorganic compounds. An inorganic compound consists of atoms that have gained or lost an electron. Such atoms are called ions. An ion that has gained an extra election has a negative charge of one and is called nonmetallic, while an ion that has lost an electron and has a positive charge of one is called metallic. Organic compounds are formed by two or more elements of the same or different elements, and they do not have an electrical charge. These compounds are united by cova-lent bonds—two atoms that share one or more pairs of electrons. An organic compound is called a molecule, and it is a stable compound. Free radicals are formed from these compounds.

Free Radicals

Two nonmetallic atoms that share a pair of electrons have a covalent bond. When this molecular bond is broken (in the case of a fire, by heat), two molecular fragments are formed, each with an unpaired electron. These molecular fragments are called free radicals. Free radicals are highly unstable and highly reactive. They are formed in a fire, and they participate in a series of chain reactions that end in the formation of more stable compounds such as water, carbon dioxide, and carbon monoxide.

Now we can explain how dry chemicals extinguish a fire. The dry chemical agent breaks down in the fire to form free radicals. These radicals unite with the free radicals formed from the fuels. This stops the chain reaction by blocking the formation of more free radicals. Thus, the fire is extinguished. This method of fire attack is known as a free radical trap, and it validates the tetrahedron theory of firefighting.

Combustion Products

The law of conservation of matter states that matter cannot be created or destroyed. Therefore, all elements or compounds that enter into the combustion process must emerge with the same elements and the same mass, though in different chemical compounds. The elements that enter the combustion process are hydrocarbon fuels and oxygen present in air. To find out what elements leave the combustion process, let's take the simplest hydrocarbon, methane (CH_4).

We will assume there is enough oxygen to burn all the methane fuel. This process is called complete combustion. The equation for the combustion of methane is

$$CH_4 + 2O_2 = CO_2 + 2H_2O$$

The symbol CH_4 stands for one methane molecule consisting of one atom of carbon and four atoms of hydrogen. When a number is shown before an atom, such as $2O_2$, the number stands for two molecules of that atom—in this case, diatomic oxygen (O_2).

Now, let's replace the element symbols with their atomic weights, shown in Table 5–1:

$$(12 + 4) + 2(32) = (12 + 32) + 2(2 + 16)$$

$$16 + 64 = 44 + 2(18)$$

$$16 + 64 = 44 + 36$$

$$80 = 80$$

Thus, 16g of methane unite with 64g of oxygen to give 44g of carbon dioxide and 36g of water (steam).

Note that four times as much oxygen is required for complete combustion of methane. This is a critical fact that increases the likelihood that not enough oxygen is available for complete combustion to occur in real–world fires. For hydrogen (H_2), each gram requires 8g of oxygen (O_2) and 34.3g of dry air. For carbon (C), each gram requires 2.66g of O_2 or 11.4g of dry air. All hydrocarbon fuels and their compounds require several times more oxygen by weight and anywhere from 10 to 30 times more air by weight. The conclusion from these facts is that for confined structure fires, fire behavior and the rate of heat release is probably limited by the amount of oxygen available.

Incomplete Combustion

If a hydrocarbon fuel consists of carbon, hydrogen, and nitrogen, then the products of combustion are carbon dioxide (CO_2), water (H_2O), and nitrogen gas (N_2). If in addition sulfur and chlorine are present, the products of combustion include hydrogen chloride (HCl) and sulfur dioxide (SO_2). In general, it is not possible to calculate the products of combustion when burning is incomplete. However, this much is known: carbon monoxide (CO) is produced. How much is produced depends upon the oxygen deficiency. For example, if 80% of the oxygen needed is present, then 10% of the combustion products by volume is CO. Carbon monoxide, of course, is a deadly gas that kills many people trapped in a fire.

Free carbon is also produced. Carbon is a unique substance in that it does not have a liquid phase. At about 6,000°F (3,315°C), carbon changes from a solid to a gas at a temperature much higher than those found in a structure fire. If abundant smoke is present, then more fire gases and carbon are being pyrolyzed (vaporized) than there is oxygen present to burn them. The fact that most of the fires that we fight involve incomplete combustion leads to a highly effective method for fighting fires with fog nozzles.

Fire Development

Once a fire starts, there are significant changes as the fire develops. These changes are identified by various fire states. They are caused by the dynamics of heat gain and heat loss. Fuels and oxygen are present everywhere, and there exists a balance between heat gain and heat loss that maintains temperature on earth within a certain range. This variation ranges from well below 0°C (32°F) to a maximum of 50°C (120°F). However, these temperatures remain well below temperatures necessary to sustain a fire at 260°C (500°F). If at a given place heat gain exceeds heat loss by unnatural means, then ignition temperatures are reached and the combustion process begins. This is the ignition state.

What happens immediately after ignition is governed by a number of things:

- The type and amount of fuel

- Fuel surface area

- Ventilation

- Structure geometry

All of these things either further fire development or inhibit it. After ignition, the fire itself generates more heat, but at the same time heat is lost through conduction, convection, and radiation. If conditions are not favorable for fire development, the fire may go out. Under certain circumstances, it may be extremely difficult to start a fire and to sustain it, as anyone who has attempted a campfire can tell you.

After the ignition state, the fire enters the growth state. In this state more and more energy is being released (heat), and the products of combustion begin to accumulate. The structure itself has little effect upon fire development since the fire is localized. This state is quite unstable. Fire may increase continuously in intensity, with an increasing rate and amount of heat released. On the other hand, the fire may slow down and enter a smoldering phase, with a decline in the rate and amount of heat released. In fact, the fire could even oscillate between these two states for a time.

If the fire continues to increase in severity, then a two–layer condition develops. The bottom part of the enclosure remains cool with near–normal temperatures and little smoke. Where the fire is burning locally, a fire plume develops. This plume consists of fire gases and normal air entrained by the rising gases. As the fire plume reaches the ceiling, a ceiling jet develops—the hot gases spread out in a circular manner, creating a hot layer of smoke and gases. This layer grows and begins to push downward as it reaches the confining walls. This state of fire development may last for only a few minutes or longer than ten minutes.

When the gas temperatures at the ceiling level reach 1,100°F (593°C), the fire is ready to enter the next state, flashover. Flashover is defined by the International Standards Organization as a rapid transition to a full surface involvement of all combustibles in an enclosure. While fire fighters may not be able to anticipate when a flashover is about to occur, there is no doubt about the fact once it has occurred. In this state, adjacent surfaces are heated together and ignite together in one rapid event. Everything in the room is suddenly on fire.

After ignition, the fuel surface available limits fire growth. As the fire progresses toward flashover, fire growth becomes limited by the amount of oxygen available, and the fire is said to be ventilation controlled. Growth is limited by the oxygen available in the enclosure or by air flowing in through vent openings. After flashover, the fire enters the fully developed stage, though fire growth is still limited by the amount of oxygen available.

At some point, a fully developed fire reaches its maximum intensity, that is, the greatest rate of heat release at the highest possible temperature. This is the fire peak, which is followed by a decline in the rate of heat release as the available fuel is consumed. In other words, the fire has entered the decay state, which may last for hours.

The peak, or maximum rate of heat transfer, occurs at the point where ventilation is just sufficient so combustion is controlled at the fuel surface. Prior to the peak at lower ventilation rates, the combustion heat release rate is less, and more unburned pyrolysis products and fuel particles are vented outside the fire in the form of smoke. In other words, it is a ventilation–controlled, or an oxygen–limited, fire. After the peak, at higher ventilation rates, more heat is removed from the fire by the excess air. This marks the decay state, with a declining rate of heat release. The important point for fire fighters is that most fire fighting is done prior to the peak, and we are fighting oxygen–limited or ventilation–controlled fires.

The Fire Environment

The houses being built in the United States today are certainly different from the houses built 50 years ago. They are better insulated, better heated and cooled, better lighted, and they have energy–efficient windows. Commercial buildings are different, too. Instead of brick and lumber, many commercial buildings are metal, or block and metal. It is generally believed that fire behavior has been affected and that fires burn hotter today compared to the 1950s or 1960s. Do the changes in the fire environment in this time interval justify this statement? One important change, of course, is the introduction and widespread use of plastics since World War II.

Plastics are made by starting with a pure chemical organic compound called a monomer. There are less than 300 monomers compared to millions of other chemical compounds. A monomer is organic since the key element is carbon. Monomers are subjected to a special chemical reaction, wherein heat and pressure cause the monomer to react with itself to form much larger carbon chains or compounds called polymers. A polymer then has other substances added to it. Thus, all plastics are mixtures. Because these mixtures are organic, they are flammable.

The nonmetallic elements found in polymers are carbon, hydrogen, oxygen, nitrogen, chlorine (Cl), fluorine (F), sulfur (S), and a small amount of silicon (Si). These substances form covalent bonds that store the energy contained in plastics. Table 5–2 lists the covalent bonds that are possible for each substance.

Table 5–2: Covalent Bonds for Nonmetallic Elements Found in Polymers

Substance	Covalent Bonds
Carbon	4
Nitrogen	3
Oxygen	2
Sulfur	2
Hydrogen	1
Chlorine	1
Fluorine	1
Silicon	4

As plastics burn, the covalent bonds are broken and their energy is released. The larger the number of bonds, the greater the amount of energy released.

However, as Frank Fire states in his book, *Combustibility of Plastics,* "This does not necessarily mean that the material with the greater amount of energy bound up in covalent bonds will burn hotter than another material. It simply means that the total heat released will be higher."[1] In other words, the rate of heat release is not necessarily greater because of the presence of plastics. Thus, there is nothing in the chemical nature of plastic compounds that will necessarily produce hotter fires.

Plastics Overview

Plastics are divided into two major groups–thermosets and thermoplastics. Thermosets are plastics that can be formed just once; any further attempts at change result in decomposition or breakdown. Thermoplastics are the larger group and contain the plastics that are most familiar, with the number–one thermoplastic by volume being polyethylene. The monomer used is ethylene (C_2H_4). By varying the temperature and pressure, many different types of polyethylene can be manufactured. Polyethylene has many different uses, such as milk jugs, film, containers, cosmetics, gasoline tanks, and pipe.

The second most common thermoplastic by volume is polyvinyl chloride (PVC). PVC is formed from the ethylene molecule by removing one hydrogen atom and replacing it with a vinyl radical and a chlorine radical. This produces the vinyl chloride monomer. PVC has more uses than any other plastic: computer parts, clothing, book covers, gloves, shoes, luggage, and pipe.

Most plastics have several additives mixed in with the monomer to provide certain properties required by the intended use. While additives may be only a small part of the plastic mixture, they have a large effect on the plastic itself and a varying effect upon the combustibility of the plastics. Additives include things such as fillers (the largest group), colorants, stabilizers, plasticizers, lubricants, and flame retardants.

Flame retardants are added to slow the rate of burning and to make the plastic more difficult to ignite. Many plastics are available with flame retardants added.

Polyethylene and its most important derivative PVC (polyvinyl chloride) belong to the group of plastics with the highest heat of combustion. Its ignition temperature is 660°F (349°C) with a heat of combustion of 43.1mJ/kg to 43.4mJ/kg (18.5Btu/lb to 18.6Btu/lb). Some forms of polyethylene are extremely difficult to ignite, primarily because of melting. PVC combustibility depends upon whether the PVC is rigid or flexible. Rigid PVC is difficult to ignite (1,035°F or 557°C) and does not produce enough heat to sustain combustion. This means that supporting flames from other substances must be used to continue burning. Flexible PVC will burn depending upon the amount of plasticizer added. Its ignition temperature is 815°F (435°C), and its combustion heat is 16.9mJ/kg (7.3Btu/lb)—about 40% of the combustion heat of polyethylene.

Plastic Combustion

Does this mean that plastics are special materials with special hazards? Do we have a different fire problem today because plastics are synthetics, not natural substances?

Frank Fire's answer may surprise you. He states that we may never know the answer. There are no simple solutions to this complex problem, that is, determining how plastics burn.

The problem is that combustibility testing of plastics is worthless as far as telling us how plastics burn in real–world accidental fires. Even testing one particular plastic tells us nothing about all of the articles that are made from this one plastic. In fact, the differences among the many kinds of plastics are as large as or larger than the differences among other groups of materials. In other words, you cannot generalize about combustibility for a single monomer. Frank Fire adds, "We cannot answer the question generally, which burns faster, plastic or wood. It seems unlikely that a correct scientific answer will ever be found because the simple question raises enormously complicated issues."[2]

So there is no scientific foundation, set of facts, or principles that can tell us how plastics burn in an actual structure fire. It is significant, however, that even with the widespread use of plastics, they form a small fraction of all the combustibles in a structure fire.

It is possible to place almost all plastics into one of four groups, classified according to how they burn. The four groups are based upon the chemical composition of the plastics:

1. Contain only carbon and hydrogen

2. Contain only carbon, hydrogen, and oxygen

3. Contain only carbon, hydrogen, and nitrogen

4. Contain only carbon, hydrogen, and a halogen (chlorine or fluorine)

The first group of plastics contains only fuels (carbon and hydrogen) and burns hotter than the other groups. The second group contains oxygen that replaces fuel in the polymer; thus, the group burns more slowly. The third group burns like the second since nitrogen replaces fuel just as oxygen does in the second group. Nitrogen does not burn; neither does oxygen. The fourth groups containing chlorine or fluorine and has the lowest heat of combustion and the lowest rate of flame spread. In fact, all of this group of plastics (with one exception) are difficult to ignite and will not burn without supporting flames from other substances. The exception is cellulose nitrate. Because of its peculiar characteristics, a cellulose nitrate fire cannot be extinguished by water.

Fighting Plastics Fires

How should fire fighters handle fires where plastics are present? Frank Fire says that such fires "...should be fought as would any other fire under the same circumstances."[3] Many believe that fires are different today from fires 40 to 50 years ago. Frank Fire agrees, but not for the reasons most people think. He says that fires are different today because the fire load is different. Frank Fire defines fire load as more than just the total mass of fuels present but "... as the total amount of energy released (total heat of combustion), its ignitibility (how easily it becomes involved in fire), and how fast the energy is released (rate of heat release)".[4] He recognizes that, because of the wide variation in the combustibility of plastics, "Today's fires are different than those 30 to 40 years ago—in that they may not be as hot burning. If occupancy fires are actually hotter today than in the past, the increased use of plastics may have little if anything to do with it."[5]

Scientific Uncertainty

Frank Fire concludes that we may never know exactly how plastics burn in structure fires. Thus, we have reached an area of scientific uncertainty. The uncertainty centers on the relationship between the fuels involved and the amount of oxygen available. However, scientists have reached one conclusion in recent years: that fire load is obsolete as a means of predicting fire behavior.

T.T. Lie states that the intensity and duration of a fire in buildings can vary by as much as 50% from the most probable prediction. He says that it is impossible to predict the time/temperature course of a fire in a building. This uncertainty is created by parameters that change with time: [6]

- Amount of fuel surface area

- Arrangement of fuels

- Wind (direction, velocity)

- Fire load

- Ventilation

- Thermal property of bounding materials

T.T. Lie states that it is only possible "…to indicate the time/temperature curve with reasonable likelihood that it will not be exceeded."[7] He does state one certainty. While it is impossible to predict what type of fire will occur in a given building, it is certain that the rate of heat release in a fire will be controlled either by the surface are of the materials that burn or by the rate of air supply through the openings. He adds,

> Usually, a ventilation–controlled fire is the more severe fire, and because of the substantial probability of its occurrence, it is common to base fire resistance requirements for building on the assumption that fire severity will be controlled by ventilation.[8]

Vyetnis Babrauskas states that the term "burning rates" is ambiguous. There are three possible meanings:

1. Initial growth rate

2. Peak rate

3. Complete time – temperature curve

Quantitatively, burning rates can be determined by the mass loss rate (kilograms per second) or by the heat release rate (kilowatts). The heat release rate has been generally accepted to be the more useful technique for measuring burning rates. Babrauskas states that techniques for measuring heat release rates were not available until a few years ago (1992), when the principle of oxygen consumption calorimetry was developed. He explains the principle as follows:

> The oxygen consumption principle states that, to within a small uncertainty band, the heat released from the combustion of any common combustible is uniquely related to the mass of oxygen removed from the combustion flow stream. The measurement technique then requires that only the flow rate, and the oxygen concentration be determined, along with the oxygen consumption constant 13.1mJ/kg of oxygen consumed.[9]

Thus, we have arrived at Thornton's rule, which is the key to determining what controls or limits the rate of heat release in a structure fire.

Thornton's Rule

The *NFPA Handbook on Fire Protection, 17th ed.,*[10] contains the heat of combustion tables for three classes of substances. Examples of those classes are included in Table 5–3.

Table 5–3: Example Heat of Combustion Values for Various Classes of Substances	
Substances	Heat of Combustion (mJ/kg) (gross)
Pure and Simple	
Carbon	32.00
Methanol	22.68
Sucrose	16.49
Plastics	
Polyethylene	46.15 to 46.2
Polystyrene	41.4 to 41.5
Nylon 6	30.1 to 31.7
Some Common Substances	
Cotton	17.5 to 20.6
Leather	18.32 to 19.8
Paper (brown)	16.3

There are some important differences in the heat of combustion of these three classes of substances. The heat of combustion of plastics ranges from 30.0mJ/kg to 47.0mJ/kg. This contrasts with the common substances, whose heat of combustion ranges from 15mJ/kg to 20mJ/kg. Other data in the table relates to the oxygen consumption constant. One is the ratio of the mass of oxygen consumed divided by the mass of the fuel, assuming complete combustion. This number (ratio) is then divided into the heat of combustion for each substance to give the ratio of the heat of combustion per kilogram of oxygen consumed.

Table 5–4 lists the ratios for pure and simple substances and plastics. There are no data for common substances.

Table 5–4: Ratios for Pure and Simple Substances and Plastics

Substances	Ratio (mJ/kg) of O2
Pure and Simple	
Benzene	13.06
Carbon	12.31
Cellulose	13.61
Ethanol	12.88
Hydrogen	16.38
Methanol	13.29
Propane	12.78
Sucrose	13.54
Plastics	
Cellulose acetate	14.67
Nylon 6	12.30
Polyethylene	12.63
Polystyrene	12.90
Polyurethane	13.66

This information is quite remarkable. There is no doubt about the conclusion drawn by Frederick B. Clark:

Examination of the heat of combustion in Appendix A [of the *NFPA Handbook*] will show that while the heat of combustion is quite different for different organic materials, the heat produced per equivalent of oxygen consumed is the same within about 10%. This fact, sometimes called Thornton's Rule, allows one to use oxygen consumption as a reasonable measure of the heat produced by a burning organic material.[11]

Let's illustrate Thornton's rule by using cellulose, the common substance of all wood products, and ethylene, a common plastic.

Cellulose $C_6H_{10}O_5$
Heat of combustion   16.12mJ/kg
Ratio O_2 mass/fuel mass   1.184

$$16.12/1.184 = 13.6\text{mJ/kg}$$

Ethylene C_2H_4
Heat of combustion   47.17mJ/kg
Ratio O_2 mass/fuel mass   3.422

$$47.17/3.422 = 13.78\text{mJ/kg}$$

Note that 13.6mJ/kg is very close to 13.78mJ/kg and that both are close to the average of 13.1mJ/kg (5622Btu/lb). This happens because ethylene requires about three times more oxygen for complete combustion compared to cellulose.

Thornton's rule is the key scientific fact used in fire engineering today. Besides that, Thornton's rule provides a solid, scientific foundation for fighting fires. Not only is the rate of heat release controlled by the amount of oxygen available, but also this is a near constant for each unit of oxygen consumed.

Thus, there has been no radical change in the nature of fires that would call for any change in strategy or tactics to cope with such changes. Thornton's Rule means that all fires are more alike than they are different. The rate of heat release will always be a near–constant 13.1mJ/kg of molecular oxygen consumed. Thornton's Rule is the single most remarkable scientific fact about fires and fire fighting.

Thus far, the experts have said that Thornton's Rule provides the most practical way to measure the rate of heat release in laboratories for both small– and large–scale testing. It is reasonable to ask this question: How about actual structure fires? Does Thornton's Rule hold there as well? Clayton Huggett of the National Bureau of Standards, Washington, DC, examined this question in detail for various fuels and products of combustion.

Since most structure fires involve incomplete combustion, the question narrows down to what effect incomplete combustion has on Thornton's Rule. Huggett reached the following conclusions.

1. The rate of heat release in a fire can be estimated with good accuracy from two simple measurements: the flow of air through the fire system and the concentration of oxygen in the exhaust system.

2. The heat release from a fire involving conventional organic fuels is 13.1kJ/g of oxygen consumed with an accuracy of ±5% or better.

3. Incomplete combustion and variation in fuel have only a minor effect on this result; appropriate corrections can be made if necessary.

4. The oxygen consumption technique of heat release measurement is adaptable to a wide range of applications, ranging from small–scale laboratory experiments to very large–scale fire system tests.[12]

Notice the third conclusion—that fire behavior in an actual structure fire has only a minor effect on Thornton's Rule. This is conclusive proof of the validity of Thornton's Rule as a scientific law that governs fire behavior in structure fires.

The conclusion from Thornton's Rule is enormous. Since structure fires involve incomplete combustion, more fire gases and free carbon are being produced than there is oxygen available to burn them. Thus, the lack of oxygen provides a real limit on the rate of heat release for a given structure fire. So no matter whether a fire is consuming hydrocarbon–based organic solids (plastics) or cellulose–based organic solids (wood), the rate of heat release is a constant 13.1mJ/kg of molecular oxygen consumed. We can safely disregard all of the variations in fuel types, geometry, heat of combustion, or any other factor of fuel load. Truly, there is no more remarkable scientific fact about fire behavior than Thornton's rule.

Window Glass

Are there any other changes in buildings that could possibly affect the nature of fires or fire fighting? There is one building feature that has changed, and it plays a crucial role in fire fighting. It is the presence, or absence, of windows. It is still true that almost every room in a dwelling has at least one window. That hasn't changed. What is different is that instead of single–pane plate glass windows, we now have double–pane windows.

First, plate glass windows (single pane) break from thermal stress early in the development of a structure fire. This begins at temperatures from 550°F to 600°F (288°C to 316°C). This temperature range is well below the temperatures needed to produce flashover at 1,100°F (593°C). This is indeed fortunate. If windows did not break until much higher temperatures, then the possibility would exist for a backdraft explosion in every structure fire. Instead, a hot, smoldering fire with backdraft potential is a rare occurrence and is limited to those structures that do not have plate glass windows in every room.

But what effect do double–pane insulated glass widows have on fire behavior? Jerry Knapp and Christian Delisio of the Rockland County, New York, Fire Training Center, conducted a side–by–side test of single–pane and double–pane windows.[13] Much to their surprise, the double–pane window broke out earlier than the single–pane window. In fact, the entire double–pane window fell out because the vinyl frame melted. Knapp and Delisio conducted a thorough search for any other data or testing done on the behavior of windows in a fire, but they were unable to find any.

However, there is some information from Robert W. Fitzgerald in the *NFPA Handbook* (17th ed.) about window glazing (glass): "It quickly cracks because of the temperature difference between the surfaces. Double–glazing does not provide much improvement. No glazing should be relied upon to remain intact in a fire."[14]

In fact, the Rockland testing detected no improvement whatsoever. Here again, there is no change that would impact fire behavior in a structure fire.

On the other hand, many commercial buildings contain wired glass, a glass sheet containing a net of steel that helps distribute heat and lowers thermal stress. This wired glass remains intact until about 1,470°F (799°C), when it begins to weaken. It will drop out at about 1,600°F (871°C). If you happen to notice wired glass in a building, watch out: a backdraft is possible.

Summary

This analysis of the scientific foundation of fire fighting may be somewhat different from what you expected because a number of myths exist in the fire service today. One myth is the idea that a change in the fire environment necessarily results in a change in fire behavior. This is true about plate glass windows. The presence of insulated glass windows has made a difference but has not necessarily resulted in hotter fires today or different ventilation characteristics. The same myth is believed with respect to the presence of plastics. Fires are not hotter today because of the change in the types of fuels consumed.

The fact is that structure fires involve incomplete combustion. In a ventilation–limited fire, the rate of heat release is a near constant no matter what the heat content of various fuels may be. Thornton's rule is the key scientific fact that, no matter what type of fuels burn, the rate of heat release will be the same for each unit of oxygen consumed in a fire

Notes

[1]Frank L. Fire, *Combustibility of Plastics* (New York: Van Nostrand Reinhold, 1991), p. 18.

[2]Ibid, preface.

[3]Ibid, p. 153.

[4]Ibid.

[5]Ibid.

[6]T.T. Lie, "Fire Temperature Time Relations", SFPE Handbook of FPE (Quincy, MA: NFPA, 1995), pp. 4 – 167.

[7]Ibid, pp. 4 – 168.

[8]Ibid.

[9]Vyetnis Babrauskas, "Burning Rates", SFPE Handbook of FPE, (Quincy, MA: NFPA, 1995), pp. 3 – 1.

[10]*Fire Protection Handbook*, 17th Ed., (Quincy, MA: NFPA, 1991), Appendix A, p. A–1

[11]Frederick B. Clark, "Fire Hazards of Materials – An Overview", *NFPA Handbook, 17th Ed.* (Quincy, MA: NFPA, 1991), pp. 3 – 15.

[12]Clayton Huggett, "Estimation of Rate of Heat Release by Means of Oxygen Consumption Measurements", *Fire and Materials* 4, No. 2 (1980), p. 64.

[13]Jerry Knapp and Christian Delisio, "Energy–Efficient Windows: Firefighter's Friend or Foe?", *Firehouse* Vol. 22 No 7 July 1977), p. 74 ff.

[14]Robert W. Fitzgerald, "Structural Integrity during Fires", *NFPA Handbook, 17th Ed.* (Quincy, MA: NFPA, 1991), pp. 6 – 7.

The Fifteen Strategic Principles

1. The control and extinguishment of an interior fire must be based upon the principle of removing the excess heat from the involved building.

2. To use water with maximum efficiency, all of it must be converted to steam.

3. The IC's size-up should determine the present fire stage and predict what and when the next fire stage will be.

4. Within a building or other confined space, the major concentration of heat is always located in the upper atmospheric level of the area of major involvement.

5. For oxygen limited fires, the expansion of water to steam deprives the fire of needed oxygen.

6. Always remember that each particular fire dictates what the water supply must be.

7. You can't put out *part* of a fire.

8. The fire will eventually burn itself out even if the fire department does nothing to stop it.

9. For a confined fire, the bulldozer attack doesn't work very well.

10. In principle, firefighting is very simple. All one needs to do is put the right amount of water in the right place and the fire is controlled.

11. It is a command responsibility to locate the center of the fire.

12. It is a command decision to choose the rate of flow for a fog attack.

13. There is no magic pill; that is, no one method of fire attack will solve all your firefighting problems.

14. If possible, it is much safer to control the fire first before sending any firefighters inside.

15. Each officer and each nozzle operator must understand fire behavior, must determine the purpose of their attack on a given day, and choose the method of attack that will fit that purpose.

Strategy is the overall plan chosen by an incident commander (IC) to combat a fire. It is derived from every aspect of the fire environment: the weather, the time of day or night, the building, preplanning, resources of both equipment and personnel, and the water supply. All of these and much more are factored into the strategic decision. Instead of broadening our analysis to consider every element that could affect the strategic decision, we want to narrow the analysis to consider only those elements that are most important when using fog nozzles. Our focus will be limited to those fires that can safely and effectively be handled by hand lines. These principles will handle almost all the firefighting done by most fire departments.

Fire Behavior

A well-trained IC must know his enemy—fire. The IC must have a thorough knowledge of what happens when uncontrolled fire begins in a structure. Occasionally, the IC may be able to shut off the oxygen supply or the fuel supply. However, most often, the IC must attack the third leg of the fire triangle—heat. This gives us *Principle One: The control and extinguishment of an interior fire must be based upon the principle of removing the excess heat from the involved building.*[1]

So the problem becomes how to remove the heat from the building. It cannot be destroyed because of the law of conservation of matter and energy. Heat is energy and, as such, the IC cannot simply destroy it by some means. So precisely how can the IC "remove the heat from the involved building" as the principle says?

We know that we use water to fight fires, and we know that water cools a fire. If a fire is cooled below the ignition temperatures of the burning fuels, then the combustion process is stopped. Applying water to a fire does not destroy the heat in any way, so what happens? What happens is called the *enthalpy of vaporization of water*. This refers to the physical process of changing liquid water to steam. This process is endothermic (heat absorbing). So the heat of the fire is transferred to the steam—it is not destroyed in this process. This leads to *Principle Two: To use water with maximum efficiency, all of it must be converted to steam.*

If enough water is applied to the fire, then the production of steam absorbs all of the heat produced by the fire. In spite of the fact that no heat is destroyed, the temperature in the fire area drops rather rapidly toward the temperature of the steam. Steam is formed at 212°F (100°C), and its temperature remains at 212°F (100°C), provided the right amount of water is used on the fire. It might seem strange that temperature drops in the fire area, yet no heat is lost when steam is formed.

The explanation is easy enough. Quite a bit of heat energy is needed to change 1lb (0.454kg) of water to steam: 971Btu/lb (2,262J/g). By contrast, it takes only 100Btu (105,500J) to raise 1lb of water 100°F (37.7°C). So it takes ten times as much heat energy to vaporize water to steam as it does to raise the temperature of liquid water 100°F. Thus, almost all of the heat produced by a fire is needed to convert water to steam. This explains how the temperature in a fire can drop rapidly to the temperature of the steam at 212°F (100°C).

We are left with a warm, moist atmosphere and no more fire. This happens because the steam temperature is well below the ignition temperatures of hydrocarbon fuels, so it is impossible for the fire to continue. The fact that steam temperature is low compared to fire temperatures means that water is the most powerful substance that can be used to fight fires. Without the enthalpy of vaporization of water, water would be a rather weak and inefficient means of fighting fires. However, no other substance comes close to water in its ability to absorb heat when it is converted to a gas. Truly, the IC has a mighty weapon at his disposal for fire combat.

To make this process work smoothly and efficiently, the IC must balance the heat-releasing (exothermic) power of the chemical reaction of the combustion process with the equally powerful heat-absorbing (endothermic) physical process of forming steam from liquid water. A balanced fire attack using the right amount of water is what the second principle calls for.

Fire is a dynamic process. That is, it is continuously changing. There is constant interaction between heat gain and heat loss. With ample fuel and oxygen available, heat gain overwhelms heat losses, and the fire increases in severity very rapidly and spreads to adjacent fuels. However, if the supply of oxygen is limited, then fire severity may decline, flaming combustion may cease, and the fire may enter a cool smoldering stage. On rare occasions, if a building is tight enough, a fire may enter a hot smoldering stage with the accumulation of flammable gases and with backdraft potential.

Depending upon the method of classification, there may be as few as three fire stages or as many as six. In any event, the IC must adhere to *Principle Three: The IC's size-up should determine the present fire stage and predict what and when the next fire stage will be.*

Principle Four is common to all fire stages as long as the fire is confined: *Within a building or other confined space, the major concentration of heat is always located in the upper atmospheric level of the area of major involvement.*[2]

This is a key element in predicting what and where the next fire stage will be. The threat of fire spread is always greater above the fire.

The rate of heat release is an important element of every fire. All authorities agree that in a structure fire the rate of heat release is controlled (limited) by either the fuel surface area available or the oxygen supply. The oxygen supply can be either within the structure or flowing into the structure from the outside.

Fires in the early flame spread stage are fuel surface limited, as are fires that are open and beyond the peak rate of heat release. Fires in between—from early flame spread on through flashover on to the peak rate of heat release—are oxygen limited. These are the structure fires that we fight most often.

Principle Five applies to these fires that are controlled by the amount of oxygen available: *For oxygen-limited fires, the expansion of water to steam deprives the fire of needed oxygen.*

One cubic foot of water (7.8gal or 29.5L) vaporized to steam will fill a small room full of steam. This is a remarkably small amount of water that can control a room-size fire. Remember that 75% of all structure fires occur in one- or two-family detached dwellings, and almost 75% of all structure fires are controlled with a single attack line and are confined to the room of origin.

This means the IC has another tremendously powerful weapon at his disposal. This blast of steam smothers the fire by depriving it of oxygen. The contaminated atmosphere produced by the fire is blasted up and out of the structure. This naturally ventilates the fire area, as a normal atmosphere quickly settles into the confined space.

So to the tremendous cooling power of steam, we can add an equally powerful smothering effect. Together, these two forces provide the IC with a firefighting capacity that can overwhelm a fire. This should give the IC confidence in his or her ability to combat the destructive forces of fire.

The Right Amount of Water

The first five principles set the stage for fire combat. The IC is armed and ready for combat. Immediately, however, the IC confronts a fundamental question: How much water is needed to control or extinguish a fire? We usually think of the amount of water needed in terms of the rate with which it is applied to a fire. This is the needed fire flow, or NFF. *Principle Six* is the key for NFF: *Always remember that each particular fire dictates what the water supply must be.* Until this fundamental fact is understood, firefighters might as well forget any other training, for it will be of little value.[3]

The word "dictates" is well chosen, and it means exactly what it says. If the NFF cannot be met, then the fire is in control. Further, the only way the IC can gain control is meeting the NFF.

Larry Davis indicates that if this principle is not understood, then all other training is practically worthless.[3] In fact, if the IC discovers that the available fire flow (AFF) is less than the NFF, then this situation makes life difficult and sometimes creates an impossible situation.

In any case, this leads to the *Principle Seven: You can't put out part of a fire.* With AFF less than NFF, the IC may decide to attack the fire anyway. However, darkening down part of the fire will not control or extinguish the rest of the fire. If you move the fire stream to another area, the fire will spread right back to the area that was just vacated. Inadequate heat removal means that some of the fuel remains above ignition temperature. As new heat is released, the inadequate water supply is boiled off, and the fire once more begins to spread. So in reality nothing has been accomplished.

Failure to control a fire when AFF is less than NFF usually results in running a race with the fire. Without immediate control, the fire has time to increase in severity. In other words, the NFF keeps on increasing. So even though the IC is able to increase the AFF, by that time the NFF is out of reach. As long as AFF is less than NFF, running a race with the fire is always a losing battle for the IC.

The only comfort in this situation is that the fire will eventually burn down to a level where AFF is greater than NFF. If not, the fire will burn out when the fuel supply is exhausted. This leads to *Principle Eight: The fire will eventually burn itself out even if the fire department does nothing to stop it.*

So fighting a fire for an extended period of time is not much of an accomplishment. This is reinforced by the realization that the insurance company will probably total the structure anyway and have it bulldozed to the ground.

Suppose that we have the opposite situation, with AFF greater than NFF. In other words, the IC has plenty of water to work with. Let's go after the fire with all that we've got. If AFF is two to three times greater than that actually needed, let's call this a bulldozer attack. Unfortunately, the IC will run headlong into *Principle Nine: For a confined fire, the bulldozer attack doesn't work very well.*

The research done at Iowa State University and by the US Naval Research Laboratory confirms that using too much water on a confined fire causes massive thermal imbalance problems. The most serious problem is endangering the safety of the attack team. Even though you won't lose control of the fire, the thermal imbalance problems are so serious that this practice cannot be recommended.

The IC doesn't have many options left. Using too little water with AFF less than NFF doesn't work. Using too much water with AFF greater than NFF doesn't work very well. This leaves only one alternative for a safe and effective fog attack: The IC must use a flow near to AFF=NFF. This ideal rate of flow restores thermal balance in the fire area and provides for a safe and effective fog attack.

Thus, we arrive at *Principle Ten* (the fundamental principle for a fog attack): *In principle, firefighting is very simple–all one needs to do is put the right amount of water in the right place and the fire is controlled.*[4]

But even Bill Nelson recognizes it is not always as simple, or easy, to put this principle into practice.

The Right Place

This fundamental principle (Principle Ten) has two elements: the right amount of water and the right place. The right place, of course, is the center of the fire, which is usually located at or near the origin of the fire. *Principle Eleven* states the importance of locating the center of the fire: *It is a command responsibility to locate the center of the fire.*

This means the IC is responsible for locating the center of the fire. If the IC delegates this responsibility to an officer, then this officer must report the location immediately to the IC when it is found.

For detached one-story dwellings, there is a simple way to locate the center of the fire. If smoke is coming out all around the eaves and the house is full of smoke with no fire visible, then the IC can circle the house and feel for the window that is hot. A hot window is a sure sign that you have located the center of the fire.

With easy access to the center of the fire, this principle can be put into practice with little difficulty. What is easy access?

1. Fire coming out one outside window or outside door locates the center of the fire and provides easy access to it. If the window does not extend down to the firefighter's shoulders, then a short ladder should be used so the fog nozzle can be operated inside the opening. Watch out for "easy" access underneath a porch or overhand— it may be dangerous.

2. Entering an outside door with air masks and full protective clothing also provides easy access. This may involve going through one room, or one room plus a hallway, to access the center of the fire.

3. Going up an inside stairway to the 2nd floor of a dwelling also provides easy access. Usually this involves going through a hallway to reach the center of the fire.

Structures other than dwellings of comparable size or height may also provide easy access no different from that of detached dwellings.

One comment should be made about attacking the center of the fire. A fog attack can be done safely and effectively from either outside or inside the structure. There is no compelling reason to insist that the only way to fight a fire is to go inside. The same scientific laws are in effect, and the same results are produced with the same attack from any point on the perimeter of a confined fire. However, it is true that an outside window can provide safer access since the opening is smaller than a doorway. Also, it provides quicker retreat from unexpected danger. So any argument about which location is safer actually favors the outside. The effectiveness of the attack, however, should be the same from either location.

What situations have difficult access? Attic fires in one- or two-story dwellings have difficult access. A favorite way to attack is to go underneath the attic fire. Every aspect of this operation is difficult. First, the ceiling must be torn open, and gravity brings the sheet rock and insulation down on top of the firefighters. Second, a folding ladder has to be placed, and there is no easy way to do this. Third, a firefighter must place the high-velocity fog nozzle into the attic space and somehow move it to hit all of the attic fire. This has to be done reaching over the firefighter's head, since the hole is not big enough for his body and air tank. None of this is easy.

An alternative way to attack an attic fire is through an outside louver. This attack is not easy, but is somewhat easier than going underneath. An opening must be laddered so a firefighter can operate the nozzle at shoulder height. Access is gained to the entire attic space for a direct attack. However, not every house has louvers, and safe access to a louver is often blocked by nearby electric lines. (Connecting an electric line next to a louver seems to be a favorite location.) A third alternative is chopping a hole in the roof. This method is the most time consuming and places firefighters in danger since they must go on the roof. Certainly, a roof ladder should always be used to provide a safe place for firefighters to stand and use as an exit if necessary. If the fire burns a hole through the roof at a vent hole, this may provide fairly easy access to the attic fire. We have seen an attic fire stopped quickly by placing a penetrating nozzle through a hole burned through by the fire.

One main problem in fighting an attic fire is the tool (nozzle) being used. A high-velocity fog nozzle is not suitable for an attic fire. A penetrating nozzle used from either above or below the attic fire should work better. We have always wondered why two Navy nozzles have not been used to fight attic fires. A short 6ft applicator with a Navy low-velocity fog nozzle on the end can be shoved through a hole in the ceiling by a firefighter standing on the floor below. This is certainly easier to execute than the current method of attack using a high-velocity nozzle.

Incidentally, Chief Lloyd Layman recommended the use of the 6ft applicator in the 1950s. Another applicator, a 12ft applicator curved at the end with an attached low-velocity Navy nozzle, also is suitable for attacking an attic fire through a hole in the roof. This requires laddering the roof and, with a firefighter standing on the ladder, placing the applicator through the hole to directly attack the attic fire.

Multi-story buildings, especially high-rises, do not provide easy access. In addition, large buildings, especially with a central hallway construction, do not provide easy access. However, with 75% of all structure fires occurring in one- or two-family detached dwelling, upwards of 75% of all structure fires should have easy access. These fires can be handled in a safe manner with minimal risk to firefighters by using hand lines. So it is only the less common structures that provide difficult access in applying the right amount of water to the right place.

Needed Fire Flow

There is ample research to justify the statement that using too little or too much water does not provide a safe or effective attack. Likewise, there is ample research to justify the statement that using the right amount of water does provide a safe and effective fog attack. The reason why the right amount of water works can be explained this way. Before the fog attack, the combustion process is in a state of thermal balance. After the fog attack, we want to have the same thermal balance, but at a much lower temperature level. A smooth transition from the hot thermal balance to the cool thermal balance is accomplished by balancing the heat releasing power of the fire with the heat-absorbing power of steam.

Since applying the right amount of water is critical to the success of a fog attack, choosing the rate of flow ranks as a command decision for the IC, which is *Principle Twelve: It is a command decision to choose the rate of flow for a fog attack.* Since the right amount of water varies as the size of the fire increases or decreases. This also implies that the rate of flow must change along with the right amount of water. In other words, greater flow rates must be used to apply larger amounts of water on larger fires.

Time

There is a fourth factor that is involved here that provides the IC with a lot of flexibility—time. Let's take a confined fire of 2,000ft³ (56,000L). Assume that its volume remains constant. This fixes the right amount of water as a constant, 10gal (37.8L). The flexibility comes in the choice of the rate of flow. This means that if a greater rate of flow is chosen, then control can be achieved in a shorter period of time. Also if a smaller rate of flow is chosen, then control can be achieved in a longer period of time. To illustrate, let's take a 2,000ft³ (56,600L) fire with 10gal (37.8L) as the right amount of water. Control can be achieved in the following ways:

Seconds (s)	30	25	20	15	10
Gallons per minute (gpm)	20	24	30	40	60
Liters per minute (Lpm)	75.7	90.8	113.5	151.5	227

Any of these combinations of time and rate of flow will provide the right amount of water for the given size fire. There are limits, however, within which the IC must operate. In other words, the flow rate chosen must be close to the ideal rate of flow for an effective fog attack.

Certainly, there is no intention to recommend that the IC attempt to do any mathematical calculations at the fire scene. There is no need to use a hydraulics calculator, a computer, or even any tables or other information on paper. In reality, the IC will be fighting fires in a room, or a series of rooms, and will need to vary the flow rate only for a small, finite number of different-size rooms. In fact, the fog nozzles available today—the ones that can vary the flow with five or six different settings—are more than adequate to fulfill the requirements of applying the right amount of water. So all the IC has to learn is how to use these nozzles effectively. The most important guideline is the realization that flow rates must be reduced well below 100gpm (378.5Lpm) for smaller room-size fires.

No Magic Pill

The statement that "there is no magic pill" is the medical version of a fundamental principle in firefighting. There are several different methods of fire attack used under different circumstances. Anyone who selects one of these methods and insists that this one method should always be used, and all others should be disregarded, is making a huge mistake. *Principle Thirteen* is the same for the fire service as for the medical service: *There is no magic pill, that is, no one method of fire attack that will solve all your firefighting problems.*[5]

Several statements advocating one method of attack are quite popular today. Two such statements follow:

- Always go inside to attack a fire

- Always attack a fire from the unburned side of a structure

It is quite easy to demonstrate that these methods cannot always be used. Take the first statement, for example. If a house is fully involved in fire, with fire coming out all windows and doors, are you going to walk inside through a door before beginning to attack the fire from inside? We don't think so, but we have seen it attempted by an unthinking IC who placed his crew in grave danger. Perhaps this not the real meaning of the statement. Perhaps what is intended is that a firefighter should attack the fire as he enters the door and continue to attack the fire from inside.

There are a couple of problems with this tactic. Sending an attack team inside necessarily means that all other operations around the structure would have to be suspended. The second problem is that water would be applied on only part of the fire at any given moment. This would run afoul of the fundamental principle that you cannot put out part of a fire. So the net result of trying to follow this method is that it unnecessarily endangers firefighters who attempt it.

We can easily demonstrate that the second statement cannot always be followed. Suppose that half of a house is on fire. The fire involves the living room, dining room, and kitchen, with fire coming out the front and back doors. Suppose further that the bedrooms are not on fire. To follow the second statement, you would have to climb into a bedroom window, go down the hallway, and start fighting the fire from there. It doesn't make sense to try to do this. We have never seen anyone attempt to do anything like that, nor do we expect to see it happen.

Principle Fourteen is related to these two statements and seems to make more sense: *If possible, it is much safer to control the fire first before sending any firefighters inside.* It is possible to gain control easily by using the right amount of water at the right place. We don't think that anyone can challenge the fundamental soundness of this principle. Safety is, after all, the first priority. Controlling the fire first is the best guarantee of a safe environment for firefighters.

Basically, firefighting is much too varied and complex for just one method of attack to work for all fire stages under any and all circumstances. While one method may be used more frequently than another, this does not justify any claim that this method should be used to the exclusion of all other methods of attack. Certainly there is no magic pill for the fire service.

The Fundamental Tactical Principle

Principle Fifteen is the fundamental tactical principle, and follows immediately from the principle that there is no magic pill: *Each officer and nozzle operator must understand fire behavior, must determine the purpose of their attack on a given day, and choose the method of attack that will best fit that purpose.*[6]

This principle involves three elements: (1) matching a given purpose with (2) a given method of attack for (3) a given fire stage. So there is much more to properly handling a fog nozzle than just opening the nozzle in the general direction of the fire.

An objection can be raised here: Isn't there really only one purpose to a fire attack—namely, to extinguish the fire? True enough, our ultimate objective is to extinguish the fire. However, this objective is different from the purpose of a given activity— say, an initial fire attack. This particular activity, which may last for a short period of time, will not necessarily reach the ultimate objective of extinguishing the fire. Many other activities will have to take place over a longer period of time before this objective can be reached. It is true that certain activities at the scene of a fire have as their purpose extinguishing the fire. But not all activities have this purpose as their goal.

One purpose, which may have nothing to do with extinguishing, is rescuing people trapped in a building. To rescue victims, firefighters may not even attempt to fight the fire. Instead, they may force their way in, clearing a path to reach the victims so they can be removed safely. On the other hand, the situation may require that the fire be attacked in order to reach the victims. So there can be more than one purpose for a given fire attack.

Notes

[1]Lloyd Layman, *Attacking and Extinguishing Interior Fires* (Quincy, MA: NFPA, 1955), p. 20.

[2]Ibid, p. 18.

[3]Larry Davis, *Rural Firefighting Operations, Vol 2* (Ashland, MA, International Society of Fire Service Instructors (ISFSI), 1986), p. 19.

[4]Floyd W. (Bill) Nelson, *Qualitative Fire Behavior* (Ashland, MA, International Society of Fire Service Instructors: (ISFSI), 1991) p. 102.

[5]John D. Wiseman, *The Iowa State Story* (Stillwater, OK: Fire Protection Publications (FPP), 1998), p. 41.

[6]Floyd W. (Bill) Nelson, op. cit., p. 107.

CHAPTER 7

Tactics

The fundamental tactical principle (Principle Fifteen) states that there are several different purposes to choose from for a given fire attack. Likewise, there are several different methods of attack that may be chosen. This chapter begins by considering three constraints (limits) that apply to any method of fog attack, whether a direct, indirect, combination, or 3–D gas cooling pulse attack. The tactics described can be adapted to any purpose or method of attack for different types of fires. These three tactical constraints follow:

1. Water must be distributed throughout the fire area. This required about ten seconds for a combination attack or a 3–D attack (several pulses). Less time may be required for smaller fires using a direct attack. This time constitutes the ideal minimum constraint.

2. The length of time that a steam blanket can hold limits the time available for an effective fog attack. This time, two to three minutes, varies depending upon the degree of ventilation of the confined fire. This limit constitutes the ideal maximum constraint.

3. Within the upper and lower constraints, if the right amount of water is exceeded, then an effective fire attack is disrupted. This variable constraint is caused by too great a rate of flow or by application for too long a time.

Constraint One

Constraint One requires that water be distributed throughout the fire area. For room–size fires, water must be distributed throughout the entire room. For smaller fires, of course, less movement is required for a direct or a 3–D pulse attack. The combination attack on a room–size fire requires the maximum amount of movement of the fog nozzle. Keith Royer and Bill Nelson, who created this attack, describe the proper distribution for a room–size fire in this way:

> The proper distribution of water fog on the initial application is very important and the nozzle operator must understand what result is trying to be accomplished. The fog pattern is adjusted so it will just reach across the area involved. The nozzle is placed inside the area and rotated following the contour of the area striking as much of the perimeter of the area as possible with the outer surface of the fog stream, across the floor, up the side across the ceiling, etc. This rotation is as violent as it is possible for the nozzle operator to make it. In placing the nozzle inside the area, it should be inside the window or other opening about an arm's length. The purpose of rotating the nozzle is to obtain steam production over as much of the area as possible. When the water strikes any of the heated surfaces in the room momentarily, it creates a steam blanket. If the nozzle rotation is rapid enough, the steam blanket will hold until the nozzle has a chance to get around to the spot again. If the rotation is too slow, the steam blanket will not hold and the fire that has been knocked down will build up before the nozzle passes its way again.[1]

Nelson and Royer strongly recommend that the rotation be clockwise from the nozzle operator's position. From their many experiments it was determined that a clockwise rotation is more effective than a counterclockwise rotation. The scientific principle that explains these differences is not known. Note that Nelson and Royer have not expressed a preference for using an outside window or an inside door. An attack from either position should work just as effectively as the other.

Constraint Two

Constraint Two implies that the fog attack is useful for confined fires only. A confined fire is highly vulnerable to an effective fog attack because it severely limits the amount of oxygen available. This does not mean that the fire must be completely enclosed. The steam blanket will hold as long as the ceiling or roof is intact, even if the fire is well ventilated. So what is a confined fire?

John A. Campbell states that the burning rate of a compartment fire after ignition is determined either by the fuel surface area available to the combustion process or by the amount of oxygen available for combustion. When a fire cannot get sufficient air to maintain the burning rate associated with fuel–surface controlled combustion, it will burn at a ventilation–controlled rate. Campbell adds,

> Considerable ventilation is required for a fully developed fire to burn at a fuel–surface rate. For example, more than one–fourth of the wall area would have to be open in a 20ft x 20ft room (6.1m x 6.1m) with an 8ft ceiling (2.4m) with an exposed combustible surface of 800ft^2 (74.3m^2) of ordinary combustibles. Many, if not most, building fires will be ventilation controlled at least during the time in which containment is a consideration.[2]

Even if a fire is burning out two windows and an interior door of a room, the fire is still ventilation controlled. As a consequence, a fog attack is useful for fighting most building fires.

There are two conditions where other tactics are indicated. The first exception occurs with fires of limited size, soon after ignition or early flame spread stage and well before the hot layer of gases banks down to the floor level. For these fires a direct attack is preferred, and it does not make any difference whether a straight stream from a fog nozzle or a solid stream from a smooth–bore nozzle is used. The US Navy's research proved that either nozzle can be used with equal effectiveness. Probably, too much water will be used on very small fires but without causing a great deal of water damage or other harm. The second exception is a steady–state fire when it breaks through the ceiling or roof. Such a fire moves steadily toward a fully open fire. In such fires steam is dispersed and dissipated so that a steam blanket does not form over the top of the fire.

Constraint Three

Constraint Three deals with the amount of water used to fight a fire. Before the attack, uncontrolled fire behavior is best described by saying that it is in thermal balance. There is a balance between the energy being provided and the energy being released. At a given height in the fire, horizontally, temperatures are equal, or in balance. For a given vertical section, temperatures continuously increase from the bottom to the top, which balances or levels temperatures. So overall, the combustion process exists in a steady state, or thermal balance.

If the right amount of water is used in a fire attack, then within a few seconds the steam condenses and appears as a white cloud that drifts upward and out of the structure. Thermal balance returns quickly once fire control is achieved. Royer describes this situation as follows:

> If at the conclusion of the knockdown the fire area is left with an even ceiling temperature of 300°F (149°C), conditions will be ideal for natural ventilation and easy and efficient overhaul. The lifting forces of the warm air (thermals) will be in balance throughout the area and we can say that we have left the area with the same thermal balance that was developed as the fire built up, but at a somewhat lower temperature.[3]

Let's summarize the rationale for a balanced fire attack. First, too little water has little effect upon the fire and no effect at all on adjacent areas. Second, the right amount of water provides the only really effective fire attack, bringing the fire under control in a short period of time. Third, using too much water is counterproductive and causes thermal imbalance. Thermal imbalance results in extreme turbulence, a disruption of the even layering and distribution of temperatures, and blocks the smooth flow of energy into and out of the fire. Visibility is destroyed. Thermal imbalance will delay overhaul, prevent the extinguishment of all of the fire, and may blow or spread products of combustion into other areas of the structure.

NFF Calculations

The general Rate–of–Flow formula may be used to calculate the NFF for varying lengths of time and varying volumes. All this is compiled in the following NFF table.

Table 7–1: NFF Table (English)

Volume (cubic feet)	Gallons	Time (seconds)				
		30	25	20	15	10
		Rate-of-Flow (gallons per minute)				
1,000	5	10	12	15	20	30
2,000	10	20	24	30	40	60
3,000	15	30	36	45	60	90
4,000	20	40	48	60	80	120
5,000	25	50	60	75	100	150
6,000	30	60	72	90	120	180
7,000	35	70	84	105	140	210
8,000	40	80	96	120	160	240
9,000	45	90	108	135	180	270
10,000	50	100	120	150	200	300
11,000	55	110	132	165	220	330
12,000	60	120	144	180	240	360
13,000	65	130	156	195	260	390
14,000	70	140	168	210	280	420
15,000	75	150	180	225	300	450
16,000	80	160	192	240	320	480
17,000	85	170	204	255	340	510
18,000	90	180	216	270	360	540
19,000	95	190	228	285	380	570
20,000	100	200	240	300	400	600

Table 7–2: NFF Table (Metric)

Volume (cubic meters)	Liters	Time (seconds) 30	25	20	15	10
		Rate-of-Flow (liters per minute)				
28	18	37	45	56	75	113
56	37	75	90	113	151	227
84	56	113	136	170	226	341
113	75	151	181	227	302	456
141	94	189	227	284	378	569
169	113	227	272	340	454	683
198	132	264	318	397	529	797
226	151	302	363	454	605	912
254	170	340	409	511	681	1,025
283	189	378	454	568	756	1,139
311	208	416	500	624	832	1,253
339	227	454	545	681	908	1,368
367	246	492	591	738	987	1,481
396	264	529	636	795	1,059	1,595
424	283	567	692	852	1,135	1,709
452	302	605	727	909	1,211	1,824
481	321	643	773	966	1,286	1,937
509	340	681	818	1,022	1,362	2,051
537	359	719	864	1,079	1,438	2,165
566	378	757	909	1,136	1,514	2,280

Now let's find out how to read and use these NFF tables. The first column on the left gives the volume from $1,000ft^3$ to $20,000ft^3$ ($28m^3$ to $566m^3$). A volume of $1,000ft^3$ ($28m^3$) corresponds to a small room, and $2,000ft^3$ ($56m^3$) corresponds to a large room. The volume of a mobile homes is around $7,000ft^3$ ($198m^3$). A small house is around $15,000ft^3$ ($424m^3$) while the average house is around $20,000ft^3$ ($566m^3$).

The second column from the left gives the number of gallons of water that is needed to fill the volume of a confined space (room) full of steam. This assumes at least 90% conversion rate to steam. Of course this is the right amount of water needed to make a balanced fire attack without creating massive thermal imbalance problems. Please note how little water is needed to control room–size fires–5gal to 10gal (18L to 37L). It is truly remarkable that most structure fires in this country can be controlled with so little water. It is also noteworthy that only 100gal (378L) is needed to control a fully involved average–size house fire. There is a huge assumption here, however, that at least 90% of the water is converted to steam within a few seconds. This is not easy to do.

Applying the right amount of water to a confined space fire can be done in many different ways by varying the rate of flow and the time involved. The remaining five columns to the right in the table give the rates of flow for different times. In other words, here you have the data used in the formula *NFF x t = gal*. For example, in row one the right amount of water is five gallons. This amount of water can be applied at a rate of 10gpm for 30s, or at 30gpm for 10s. In the first instance, flowing 10gpm for 30s certainly applies 5gal of water to the fire. In the second instance, flowing 30gpm for 10s also applies 5gal to the fire. The same numerical relation extends throughout the table for each row.

Each column gives a rate of flow for a fixed time. The left hand column has a fixed time of 30s. As the rate of flow increases, more water is applied to the fire. More water means that bigger fires can be controlled within 30s. This is the time used in the Iowa Rate–of–Flow formula. The research done at Iowa State University showed that almost all fires could be controlled within 30s. This is the longest time that should be used for an initial fog attack. Since a fog blanket will hold for only 2min to 3min, this is the time that you have to enter the structure and completely extinguish the fire. Overhaul of the fire building will ake considerably longer, of course.

The right hand column is for a fixed time of 10s. The is the minimum time that should be used for a fog attack so that you can distribute the water throughout the fire area. Notice the first two entries in the 10s column. Rates of flow from 30gpm to 60gpm (113Lpm to 227Lpm) should be used for room size fires. Standard flows from preconnected attack lines are much too great to handle such small fires. Standard flow from a 1.5in (38mm) attack line of 100gpm (378Lpm) is too great for a large room of 2,000ft³ (56.6m³) and much too great for smaller rooms. Standard flows, quite simply, do not give you enough time to properly distribute the water throughout the room.

As an example, let's take a small room of 1,000ft³ (28.3m³) with a flow rate of 100gpm (378Lpm):

$$100 \times t = \frac{1000}{200} \qquad 378 \times t = \frac{28.3}{1.5}$$

$$100 \times t = 5 \qquad\qquad 378 \times t = 18.8$$

$$t = 5/100 \qquad\qquad t = 5/100$$

The fraction 5/100 equals 1/20 of a minute, or 3s. It is impossible to distribute 5gal (19L) of water equally throughout a confined space in 3s. If the nozzle is kept open for longer than 3s, then thermal imbalance is created and an effective fog attack is disrupted.

The situation is much worse for a 1.75in (44mm) attack line or for the 2in (51mm) attack line. Two 1.5in (38mm) lines are operated together in the same confined space are equivalent to using a single 2in (51mm) attack line.

So the NFF tables illustrate nicely what is required to make an effective fog attack. For smaller rooms, the largest flows that should ever be used are 30gpm to 60gpm (113Lpm to 227Lpm). This is the critical information that the IC needs to know when sizing up a structure fire. Of course, smaller flows may be used, but this just lengthens the time needed to control these room–size fires. We don't think anyone is advocating a return to using booster lines for initial fire attack. It is much better to provide the larger flow of 30gpm to 60gpm (113Lpm to 227Lpm) since this shortens control time to the minimum possible for a fog attack. If you can do it in a shorter time, why not do so?

Let's take a look at the opposite corner of the NFF tables with flows of 30s and a volume of 15,000ft³ (424.5m³), a small house, and a volume of 20,000ft³ (566m³), an average–size house. Even if the flows are adequate, we are faced with an entirely different situation than that of the one–room fire. It is simply not possible to fight a fire from one position if the fire is burning throughout the entire house. Make no mistake about it: water must be distributed throughout the fire area.

What is needed, for sure, are multiple attack lines. Standard operating procedure for any structure fire is to cover all four sides. This means four attack lines, one on each side of the house. A flow of 60gpm (227Lpm) from each line can control a small house fire in 15s to 20s. Two pumpers are needed for this, with each truck supplying two preconnects. We have seen a four–line attack operate with spectacular results. All four lines were opened up at the same time, and the fire was blacked out immediately throughout the entire house. Note that with four attack lines, flows need to be reduced below 100gpm (378.5Lpm). In effect, you are making four separate room–size attacks in different parts of the house.

Let's conclude the NFF calculations with an actual house fire. This fire occurred in February, 1991, in a small, square four–room house. There were two bedrooms on the left side (B side), one behind the other. On the right side (D side) there was a living room with a kitchen behind. A small bathroom was to the right off the kitchen.

When the first truck arrived, the fire was burning out the front door, the right front window (A side), and the first window on the D side. All of these opened into the living room. A mass of flames enveloped the small porch at the front door. Later, we found out that a 2gal can of kerosene sitting just inside the front door had exploded. The fire had spread into the kitchen at the ceiling level and had almost burned through the wood–panel bathroom door at the top. The front bedroom was scorched at the ceiling level. The rear bedroom had some smoke damage.

A combination attack was made through the living room window at the D side with a 1.75in line. The nozzle was opened to detent stop three (TFT automatic). The nozzle was rotated inside the window two complete turns. Almost immediately, the flames changed to white condensing steam. The nozzle operator backed away from the window as condensing steam spread out the window and enveloped the entire house. Much to our surprise, the flames outside the house at the front door were extinguished, too. An officer called the nozzle operator around to the front, and a straight stream was used to hit burning embers (red hot) above the front window. Water was never applied in the front bedroom or in the kitchen.

The nozzle was open halfway (detent stop three) for about 10s. The estimated flow rate was 75gpm (284Lpm). Using the rate formula, NFF × t = gal

$$75 \times 0.16 = 12.5\text{gal} \qquad 84 \times 0.16 = 45\text{L}$$

The decimal 0.16 is equivalent to 1/6 of a minute, or 10s. The 12.5gal (45L) was water that produced the following volume of steam:

$$12.5 = \frac{\text{Vol}}{200} \qquad\qquad 284 = \frac{\text{Vol}}{1.5}$$

$$200 \times 12.5 = \text{Vol} \qquad 284 \times 1.5 = \text{Vol}$$

$$\text{Vol} = 2{,}500\text{ft}^3 \qquad \text{Vol} = 4.26\text{L}$$

Since the numbers for time and flow rate are both estimates, it might be better to say that approximately 15gal (56.7L) of water was used on this fire. Fifteen gallons (56.7L) are enough to fill 3,000ft^3 (84.9m^3) full of steam.

The actual volume of the three rooms that were involved in fire (excluding the rear bedroom) was 2,750ft^3 (77.8m^3). That number is between 2,500ft^3 and 3,000ft^3 (70.75m^3 and 84.9m^3). So the NFF calculation turns out to be highly accurate—accurate enough to verify that the fog attack was close enough to the right amount of water needed for fire control.

Incidentally, the living room was cleaned out completely, and less than 50gal (189L) of water were applied to smoldering debris outside the house. The combination attack in this fire came very close to extinguishing the fire. Since there was little structural damage to the house, it was rebuilt inside and is still in use today.

The Art of Firefighting

The art of firefighting relies upon the judgment and experience of the nozzle operator and the officer in command. This art demands more than just opening up a fog nozzle wide open and blasting away at the fire—a bulldozer approach that is counterproductive and disruptive, causing thermal imbalance and unnecessary water damage. Before flowing water, the "artist" must stop a moment and think.

When initiating a fog attack—whether a direct, a 3–D pulse, or a combination attack—a firefighter must determine if the available fire flow meets the NFF for fire control of a confined fire. Once the point of attack is chosen, the nozzle operator (and officer) must decide on four things: .

- **Adjust the fog nozzle to provide the proper flow.** Certainly for one– or two–room fires, the nozzle operator should select the first or second detent stop or the smaller flows on the volume control ring.

- **Adjust the fog pattern to provide the proper reach.** The pattern must be adjusted so the fire stream just reaches across the fire area. A straight stream with a 50ft reach is too great for a one–room fire.

- **Distribute the water properly throughout the fire area.** The most effective way to do this for a room–size fire is by a combination attack with a clockwise rotation of the nozzle placed just inside the opening to the fire area.

- **Shut down the nozzle at the proper time.** The operator needs a reasonable expectation of when fire control will be achieve. He must shut down the nozzle when flames are blacked out and white condensing steam appears and begins to drift up and out of the structure.

There is much more to firefighting than just opening and shutting a nozzle aimed in the general direction of the fire. The art of firefighting using a balanced fire attack requires a rather careful, thoughtful handling of the nozzle. Essentially, you are

Table 7–3: Task Force Tips Automatic Nozzle Data (English)

Length (ft)	Pump (psi)	Performance at Detent Positions, 1¾ hose, 100psi nozzle					
		6	5	4	3	2	1
		(Flows in gallons per minute rounded to nearest 5gpm)					
150	125	110	105	100	80	55	35
150	150	145	140	125	95	65	45
150	175	175	170	155	115	75	50
150	200	210	200	175	130	80	50
200	125	100	95	90	75	50	25
200	150	130	125	115	85	55	45
200	175	155	150	140	110	75	50
200	200	180	170	160	125	80	50
300	125	85	85	80	70	50	35
300	150	110	105	100	85	60	45
300	175	125	125	120	100	70	50
300	200	145	145	135	110	80	50

Table 7–4: Task Force Tips Automatic Nozzle Data (Metric)

Length (m)	Pump (bar)	Performance at Detent Positions, 44mm hose, 6.9bar nozzle					
		6	5	4	3	2	1
		(Flows in liters per minute rounded to nearest Lpm)					
46	8.6	416	397	378	303	203	132
46	10	549	530	473	359	246	170
46	12	662	643	586	435	284	189
46	14	795	757	662	492	303	189
61	8.6	378	360	340	284	189	95
61	10	492	473	435	322	203	170
61	12	586	567	530	416	234	189
61	14	681	643	605	473	303	189
91	8.6	321	321	302	265	189	132
91	10	416	397	378	321	227	170
91	12	473	473	454	378	265	189
91	14	549	549	511	416	303	189

balancing the power of steam (endothermic) with the power of fire (exothermic). The artist is capable of achieving fire control in the least possible time, with the minimum amount of water, and with a minimum of risk.

The solution lies in using two types of fog nozzles that are available today: the variable–flow nozzle with manual volume control and the automatic nozzle with detent stops. Tables 7–3, 7–4, 7–5 and 7–6 provide information about three nozzles manufactured by Task Force Tips, Akron Brass, and Elkhart Brass. The information provided is the pump pressure, hose length, nozzle pressure, and flow rate for each setting of the nozzle.

Each of the nozzles described in Tables 7–3, 7–4, 7–5, and 7–6 is capable of reducing the flows below what is considered standard for the attack line used so that the ideal rate of flow can be used for room–size fires. Interestingly, very little water is needed to control the typical structure fire, that is, a one–room fire in a one– or two–family dwelling. The General Rate–of–Flow formula is the only formula that can determine the right amount

Table 7–5: Elkhart Brass Select–o–Flow Nozzle Data

Discharge in gallons per minute

Setting	Nozzle Pressure — pounds per square inch (top row)							
(gpm)	40	50	75	100	125	150	175	200
40	25	29	·34	40	44	48	52	56
60	39	44	54	60	70	77	85	91
95	59	67	82	95	105	114	123	132
125	79	88	105	125	140	154	168	182

Discharge in liters per minute

Setting	Nozzle Pressure — bars (top row)							
(Lpm)	2.8	3.4	5.2	6.9	8.6	10.3	12	13.8
151	95	110	129	151	166	182	197	212
227	148	166	204	227	265	291	322	344
360	223	254	235	360	397	431	473	500
473	299	333	397	473	530	583	598	689

Table 7–6: Akron Brass Turbojet Nozzle Data

Nozzle Pressure (psi) (bar)		Flow Setting (gpm) (Lpm)		Actual Flow (gpm) (Lpm)	
75	5.0	30	115	26	98
		60	230	52	197
		95	360	80	310
		125	475	125	475
100	7.0	30	115	30	115
		60	230	60	230
		95	360	95	360
		125	475	125	475
125	8.5	30	115	34	129
		60	230	67	254
		95	360	106	401
		125	475	140	530

of water for such fires. Thornton's Rule (see chapter 5) states that the heat release rate for hydrocarbon fuels is 13.1mJ/kg of molecular oxygen consumed. This near constant is the foundation for really effective firefighting using a balanced fire attack.

Fog Tactics

A question that generates one of the most heated debates in the fire service is, "What is the best method of attacking fire in structures?" The question itself is the cause of the controversy, since firefighting is much too varied and complex for one method of attack to be able to solve all of our firefighting problems. Instead, the question that should be asked is based upon the guiding principle as stated by Bill Nelson: "Each officer and nozzle operator must understand fire behavior, must determine the purpose of their attack on a given day, and choose the method of attack that will best fit that purpose."[4]

This principle implies there is no one best method of attack that will fit all purposes or all types of fires. Before opening up a nozzle, you must make a choice. So the question that should be asked is this: What method of attack will best fit a given purpose for a give type of fire?

Purposes

While it is true that one purpose of a fire attack is to extinguish the fire, this usually occurs with smaller fires in the beginning stage of development. Even in this case, a situation may arise where direct access to the fire is blocked. This requires a different method of attack. Also, if a fire attack would scatter burning materials, a different method of attack will be required. So even in the simplest situation where the purpose is to extinguish the fire, different methods of attack may have to be used.

Not always is the purpose of an attack to extinguish the fire. In a multistory building with a fire burning on the 1st floor, a dangerous situation is created. Flammable gases can accumulate in the upper floors with a lack of oxygen for flaming combustion. In such a situation, extinguishing the fire on the 1st floor is the wrong thing to do. What should be done is to produce quantities of steam over a period of time—say, several minutes. The steam will move into the upper floors, condense, and cool the area, thereby preventing the spread of the fire to the upper floors. So the purpose of this attack is not to extinguish the fire but to use the fire to produce the steam necessary to eliminate a dangerous situation on the upper floors—a holding action.

Another purpose is to control or knock down the fire. Control does not mean extinguishing the fire. In fact, after an initial attack on a well–developed fire, if nothing is done after the initial attack, the fire will redevelop. This occurs because of the accumulation of char. In other words, the initial attack does not usually extinguish the fire. Further action is required to extinguish the fire.

Another situation occurs rarely; but when it does occur, it is the first priority at any fire. This is a rescue situation. In this case, the purpose is to go in and rescue (or recover) victims trapped in a burning building. Extinguishing or controlling the fire may be completely disregarded. At least, this is not the primary purpose of the attack. Firefighters want to force their way in, secure the victims, and bring them outside as quickly as possible. Of course, attacking the fire may be the best method for doing this. However, the method for making a rescue is usually different from the methods of attack in a non–rescue situation.

Finally, for the largest fires that a fire department may encounter, it may not be possible to control or extinguish the fire in a reasonable length of time. The purpose in this case is to stop the spread of the fire either within the building or to other buildings. This purpose may also be called protecting exposures, a defensive operation.

Methods of Attack

Now let's turn to the methods of fire attack. Three methods should be familiar to all firefighters: the direct attack, the indirect attack, and the combination attack. The direct attack, of course, applies water directly to the burning surfaces. The indirect attack does not apply water directly to the burning surfaces; instead, water is applied into the heated overhead of a confined space. The combination attack is a combination of a direct and an indirect attack. A fourth attack, created in Europe, is the 3–D pulsed gas cooling attack, which differs from the indirect attack when it is applied overhead in a confined space.

Bill Nelson in his book Qualitative Fire Behavior introduces several other names for different methods of attack:

- SSOC attack – straight stream off the ceiling

- WIN attack – applying water through a 2nd floor window

- Bulldozer attack – grouping several lines together

- Deluge attack – multiple master streams and hand lines

Following are illustrations of the use of the eight different methods of fire attack mentioned above. The choice of the best method of attack is guided by applicable research, indicated at the end of each illustration.

1—Direct attack. Purpose: extinguish during early flame spread (stage 1A). Water is applied directly to the base of the flames, or to the "seat" of the fire. A straight stream from a fog nozzle or a solid–stream nozzle is usually big enough to cover the entire fire area with little movement of the nozzle. A wider fog pattern may be used, but it must have sufficient reach to hit the burning surfaces directly. A short burst is almost always sufficient to extinguish the fire. (Research Source: Chief Layman, US Naval Research Laboratory)

> *a. Direct attack.* Purpose: extinguish during ignition or flame spread (stages 1A, 2A). If the fire involves a wider area in a room, the direct attack cannot be made by pointing the nozzle at the base of the flames. The water must hit directly all burning surfaces. The nozzle must be moved in a 180° arc centered at the nozzle. This is a direct 180° attack.

2—SSOC (straight stream off the ceiling). Purpose: extinguish during ignition, flame spread (stages 1A, 2A). If contents or panels block direct access, then a different method of attack must be used. The idea behind SSOC is to hit the ceiling so the water spreads along the ceiling and falls down on the fire. This is not an indirect attack but is actually a direct attack. The nozzle may have to be moved to cover the entire fire area. SSOC may also be used when you wish to avoid scattering loose materials if you hit them directly with a straight stream. The spray off the ceiling would not have enough force to scatter anything. In this case, the overhead presents no threat to the firefighters. (Research Source: Floyd W. Nelson)

3—3-D Pulse attack. Purpose: control during flame spread, cool smoldering (stages 2A, 2B). If the fire has progressed so that the heated overhead presents a threat to firefighters, then the 3–D pulse attack should be used. First, the overhead must be cooled and the threat of flashover removed. Then, the firefighters can safely continue to move into the room and extinguish the fire by a direct attack. (Research Source: US Naval Research Laboratory, European Research)

4—Indirect attack. Purpose: hold during flame spread, steady state (stages 2A, 5). In a multistory building with fire on the 1st floor and flammable gases accumulating in the upper floors, the purpose of the attack is to engage in a holding action. The fog pattern must be adjusted so it does not hit any walls or the ceiling. Also, water must not be applied directly to the fire itself. The fog pattern must be carefully worked back and forth at the ceiling level to obtain the maximum steam production. The steam will dilute the flammable gases on the upper floors, condense, and cool the area. This will stop the spread of fire to the upper floors. After this happens, the 1st floor fire can be extinguished by any appropriate method.

This tactic is very similar to the European 3–D pulse tactic. The fog does not hit the ceiling or walls, or even the fire itself. Pulse tactics could be used to accomplish this. However, the purpose of Nelson's tactic is to maximize steam production, whereas the 3–D pulse tactic is designed to minimize steam production. So this purpose is fundamentally different from the European tactic. (Research Source: Floyd W. Nelson)

5—WIN attack. Purpose: hold during flame spread, steady state (stages 2A, 5). If a fire is burning on the 2nd floor and close access is not possible from either inside or outside, then the SSOC attack can be modified to achieve a holding action. Standing a short distance from the building, the nozzle operator must move from one edge of the room to the other in a circle or straight line, simultaneously pouring water into the window. While this may not achieve complete coverage of a 180° direct attack, it can come close. This should be quite effective as a holding action. (Research Source: Floyd W. Nelson)

6—Combination attack. Purpose: control during cool smoldering, hot smoldering, steady state (stages 2A, 4, 5, 6). If a fire fully involves a room or a confined space or is a smoldering fire, the purpose of the attack is to knock down the fire in the shortest possible time. The four tactical guidelines should be followed.

> a. *Progressive attack*—Purpose: control during flame spread, hot smoldering, steady state, clear burning (Stages 2A, 4, 5, 6). When a fire involves multiple rooms of a structure and not enough lines are available to attack each room at the same time, a progressive attack may be used. This not a new method of attack. Rather, it is using a given method of attack in sequence, one room after the other. (Research Source: Keith Royer and Floyd W. Nelson)

7—Bulldozer attack. Purpose: rescue during any fire stage. While it rarely occurs, a rescue situation deserves top priority for any fire department. If victims are trapped in a building, the purpose is to make a rescue (or recovery) and bring the victims outside as quickly as possible. In the bulldozer attack you want to force your way in with as much water as possible, protecting the rescuers, removing the victims, and backing out as quickly as possible. This method of attack groups two or more lines together. One of the lines may be needed to protect the entrance, another line to protect a hallway, and a third line to go with the rescue team. This concentration of lines is counterproductive in non–rescue situations. (Research Source: Floyd W. Nelson)

8—Deluge attack. Purpose: stop spread during steady state or clear burning (stages 5, 6). This method of attack is used when the fire department is unable to gain control within a reasonable length of time. These are larger fires, and the operations are usually defensive. The purpose of the attack is to prevent the spread of the fire within the building or to other buildings.

The deluge attack is the familiar surround–and–drown tactic. This attack uses large master streams as well as multiple hand lines. Direct access may not be possible because of the danger of collapse. Nelson has an apt description of these operations. He says that all kinds and sizes of streams are put into operation and simply throw water in the general direction of the building. This type of attack takes a bit longer because the fire must burn though the roof before it can get at the water.[5] (No research source)

These ten situations illustrate the tactical decisions that must be made on the fire ground. Create your own tactics that suit your department. Choose a fire stage in a typical building, determine the purpose of your attack, and select the method of attack that best fits that purpose. For example, create tactics for a blitz attack. Also as a simple exercise, take a rectangular structure that is burning on the ground. Devise tactics for coordinating two attack lines, beginning by attacking along the two diagonals of the rectangle, and continue a direct attack until all of the burning surfaces are cooled. As long as you are guided by the research on fog nozzles, you should be able to mount a fog attack safely and effectively.

Methods to Avoid

Let's consider two methods of attack that do not work. The first method is called "outstanding firefighting"—a name certainly not intended to be complimentary. The authors first heard this term used by Mike McVey, instructor at Tennessee State Fire School. It means standing still some distance outside the structure and projecting a straight stream through the nearest available opening. Water from a straight stream moving at 98ft/s (68mph or 109km/h) travels through a fire area 14ft (4.3m) in 3/10 of a second. Very little of this water is evaporated by the fire. The result is an almost complete waste of water. Bill Nelson describes this as follows: "The fire will set up convection currents around the fire stream and burn merrily on."[6] "Merrily" clearly indicates that the fire is in control and that the method of attack is not working. The error, of course, is the failure to make any attempt to distribute the water throughout the fire area. In fact, it would be impossible to do this standing some distance from the building.

The second method is a wide–angle fog attack. This means opening the nozzle to at least a 60° angle as you advance into the fire area. With the short reach of the stream and the amount of air entrained, the result is the creation of extreme turbulence that upsets the thermal balance. In other words, this attack may very well push the fire around. These problems are compounded by the fact that entirely too much water may be used. Thus, we have a bulldozer attack using a single line. This is the method Nelson complained about when he used the phrase "dark decade of thermal imbalance".[7]

Gradually, fire departments have learned that instead of pushing the fire inside, if they attack from the opposite direction, then they can push the fire outside. This leads to the principle that the fire must be attacked from the unburned side of the structure. Thus, today's preferred method is making a direct attack with a straight stream.

Many fire departments consider this to be the best method of attack and, hence, the only method of attack really needed by any fire department. However, this opinion is based upon a misunderstanding and misapplication of scientific principles. The origin of this method of attack lies in the misuse of fog nozzles that occurred in the 1960s and 1970s, the wide–angle attack coupled with outstanding firefighting. The problems encountered led to the creation of new tactics requiring artificial ventilation to compensate for the use of too much water.

By contrast, a combination attack using the right amount of water does not require artificial ventilation since natural ventilation by steam removes the heat from the fire area. Even so, the combination attack will not work for all fires. In fact, there is no one best method of fire attack that will work for all fires. All methods have their place in the arsenal of methods of fire attack. It all depends upon the purpose of the attack, the type of fire, and the best method for that purpose.

One aspect of the direct attack needs to be addressed. If fire department policy is to always make an interior attack, it will lead to a dangerous situation in certain cases. For non–routine fires, this method places firefighters at grave risk of injury or death by collapse or flashover. The proliferation of truss roofs has further increased the danger to firefighters who make an interior attack. Too many times, tragic incidents are reported of needless deaths or injury caused by over reliance upon one method of fire attack.

Bill Nelson's guiding principle, choosing the best method for a given purpose at a given fire, certainly includes careful consideration of the safety of the firefighters involved. If firefighters realize they have a choice of methods of attack, then surely they can choose the method that best ensures their safety and survival on the fire ground.

The realization that fog nozzles can be used safely and effectively using different methods of attack should decrease the risk that firefighters face on the fire ground. However, for this to be done, a drastic change in attitude of fire commanders will be required. This needed change has been well expressed by Ed Comeau:

> According to NFPA 1500, Fire Department Occupational Safety and Health Program, the incident commander is required to integrate risk management into the regular functions of incident command. He or she must thus limit aggressive firefighting to situation where lives are endangered and can possibly be saved, which means reducing risks to firefighters operating to protect property only. The standard goes so far as to say that no risk to firefighters' safety is acceptable when there's no possibility of saving lives or property.
>
> The incident commander is also charged with evaluating the risk to members in terms of the purpose and potential results of their actions in each situation. Where the risk to firefighters is excessive, the standard calls for use of defensive operations only. And when fire involves a wood truss, the risk is compounded by the fact that flames may stay hidden inside the truss structure, taking firefighters by surprise when the roof or floor fails.
>
> It is vital that incident commanders placing firefighters in hazardous situations ask themselves one fundamental question: "What are we trying to accomplish?" If lives can be saved, then calculated risks may be taken. If the building and its contents are the only things in danger, the fire ground strategy must take this into account. Incident commanders with qualms about taking a less aggressive approach should ask themselves whether they should put their firefighters at risk for a building owner who hasn't protected his or her property with a sprinkler system. Why risk irreplaceable lives to save replaceable property?[8]

There should be no doubt as to the correct answer to this last question.

Notes

[1] Report of Story City Fire Test, Firemanship Training, Engineering Extension Service, Iowa State College, Ames, IA, August 29, 1954.

[2] John A. Campbell, "Confinement of Fires in Buildings", *NFPA Handbook, 17th Edition* (Quincy, MA: NFPA, 1991), pp. 6 – 80.

[3] Keith Royer, "Water for Fire Fighting", Iowa State University Engineering Extension Service Bulletin 18, (1959), p. 22.

[4] Floyd W. (Bill). Nelson, *Qualitative Fire Behavior* (Ashland, MA: IFSTA, 1991), p. 109.

[5] Ibid.

[6] Ibid, p. 100.

[7] Ibid.

[8] Ed Comeau, "Roof Collapse Kills Three" NFPA Journal, Vol 93, No. 4 July–August 1999, p. 80.

CHAPTER 8

The Mechanics of Fog Nozzles

Throughout most of mankind's history, firefighting was done by manually hauling water in buckets and throwing it on the fire. There were no fire pumps, fire hoses, or nozzles. In the 1700s, manually operated pumps, copper-riveted leather hose, and copper/brass nozzles came into use. By the middle of the 1800s, the steam fire engine provided the first real revolution in firefighting. Together with cotton fire hose and smooth-bore (solid-stream) nozzles, firefighters for the first time had a real chance to control and extinguish fires.

Smooth-bore nozzles were used exclusively by fire departments for a hundred years, from about 1850 to 1950. But by 1950, new fog nozzles had been invented that were quite successful in fighting Class B fires. Eventually, almost all fire departments adopted fog nozzles as their primary tool in fighting Class A as well as Class B fires. This almost universal shift from one primary nozzle type to another constituted a major revolution in firefighting. So why was this shift made?

The answer, quite simply, is that fog nozzles are better than smooth-bore nozzles. So let's compare these nozzles to find out why.

What is a Nozzle?

A nozzle is a device attached to the end of a fire hose that is designed to produce a useful fire stream. A nozzle creates such a stream in four ways:

1. Controls the flow

2. Provides reach

3. Shapes the stream

4. Determines direction[1]

All nozzles create a fire stream in the same basic way–by decreasing the size of the nozzle outlet (tip) compared to the size of the hose. For example, the diameter of the circular tip of a smooth-bore nozzle must be no greater than one-half the diameter of the hose. (See Figure 8–1) A smooth-bore nozzle controls the flow by the size of the tip. A given tip with a given pressure at the nozzle and a given size hose provides a single rate of flow. To change the rate of flow significantly, a different size tip must be used.

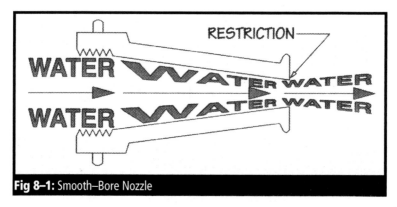

Fig 8–1: Smooth–Bore Nozzle

Squeezing water through the tip increases the pressure at the nozzle, increasing the speed (velocity) of the water. The speed determines how far the water is projected (reach). The effective reach of a solid stream at 50psi (3.47bar) nozzle pressure varies from 73ft (22m) vertical and 61ft (18.6m) horizontal for a 1in (25.4mm) tip up to 79ft (24m) vertical and 75ft (22.8m) horizontal for a 1.5in (38mm) tip. The maximum reach is obtained by holding the nozzle at a 30° angle to the ground.

The shape of the smooth-bore stream is a 3-D cylinder. However, this solid stream does not remain solid for long. The stream begins to expand just as soon as it leaves the nozzle. Further, friction with the surrounding air, caused by piling up and displacement, begins to tear away at the stream. Also, gravity pulls the stream down.

John R. Freeman, creator of the Underwriter's Playpipe, the most efficient solid-stream nozzle (1888), defined an effective solid stream as follows:

1. At the limit named has not lost continuity of stream by breaking into showers of spray.

2. Up to the limit named appears to discharge 9/10 of its volume of water inside a circle 15in (38cm) in diameter and $^3/_4$ of it inside a 10in (35.4cm) circle .

3. Is stiff enough to attain in a fair condition the height or distance named even though a fresh breeze is blowing.

4. At a limit named will, with no wind, enter a room through a window opening and just barely strike the ceiling with force enough to splatter well.[2]

This definition is still used today with minor modifications.

Fog Nozzles

By 1951, two types of fog nozzles were in use. The Navy nozzles produced a spray by impinging jets of water, that is, the jets collided, thereby breaking water into smaller drops. The industrial nozzles produced spray by peripheral deflection, that is, a baffle centered in the nozzle deflected water around the circumference of the circular tip. (See Figure 8–2)

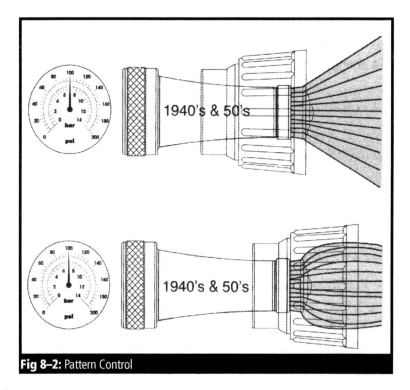

Fig 8–2: Pattern Control

The first major improvement in fog nozzles came in the 1950s when Keith Royer and Bill Nelson of Iowa State University saw the need to keep the flow constant no matter what the shape of the fog stream They cooperated with Akron Brass Company in producing the first constant flow fog nozzle. The size of the tip opening was kept constant by inserting a second internal collar (or floating stem). The constant flow nozzle provided the right tool for the Iowa Rate-of-Flow formula.

The second major improvement in fog nozzles came in 1969 when Chief Clyde MacMillan, Commandant of the Gary, Indiana, Fire Task Force, saw the need for some mechanism whereby the fog nozzle could change the flow rate without having to change tips or even nozzles. He designed a hydraulically controlled device that could be added to existing high-volume fog master stream nozzles. MacMillan formed his own manufacturing company, Task Force Tips, because existing nozzle manufacturers were not interested in the new nozzle (See Figure 8–3). However, two years later, Elkhart Brass Company produced a similar master stream nozzle.

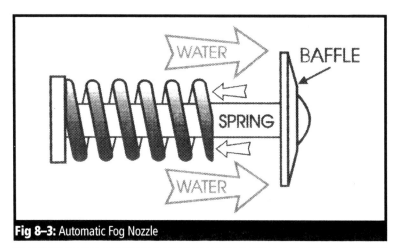

Fig 8–3: Automatic Fog Nozzle

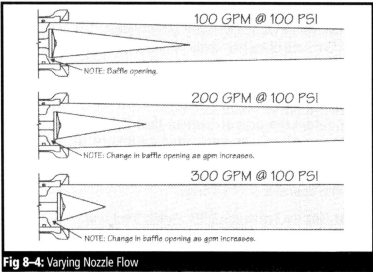

Fig 8–4: Varying Nozzle Flow

This device consisted of a coil spring wrapped around the stem of the baffle. Thus was born the automatic (constant pressure) fog nozzle (Figure 8-4).

The automatic nozzle uses a principle very similar to that of a pumper relief valve. The pressure control mechanism senses the pressure at the base of the stem. Slight adjustments are made automatically to maintain the optimum nozzle pressure, 100psi (6.9bar). The baffle attached to the pressure control unit varies the discharge opening of the nozzle. In effect, the nozzle is continuously changing "tip" size to match the water being delivered.

A comparison may help explain how automatic nozzles work. A smooth-bore (fixed gallonage) nozzle is similar to using a manual transmission on a car. As the vehicle speed (flow) increases or decreases, the correct gear (nozzle opening) must be selected by the driver (nozzle operator) and manually changed for proper speed (stream quality).

The automatic nozzle is similar to an automatic transmission. As the vehicle speed (flow) increases or decreases, the correct gear (nozzle opening) is automatically selected, producing the proper speed (stream quality) all of the time. For whatever "speed" (flow) you choose, the automatic nozzle will adjust to give you the proper "gear" (flow opening).[6]

Smooth-Bore vs. Fog Nozzles

The following is a comparison of the two types of nozzles:

Fog Nozzle	Smooth Bore Nozzle
Breaks water into little drops	Does not
Shape and reach vary	Fixed narrow stream
75psi or 100psi (5.2bar to 6.9bar)	50psi (3.47bar) hand lines
	80psi (5.5bar) master streams
Higher pump pressures	Lower pump pressures
Entrains air	Does not

Using this comparison, let's begin with an important principle of firefighting. Lloyd Layman stated this principle as follows: "The control and extinguishment of interior fires must be based upon the principle of removing excessive heat from the building."[7] For an effective fire attack, we therefore need the best heat-absorbing capacity that water can provide. Increasing the surface area of the water does this. The rate of heat transfer is proportional to the surface area of the water. The best range of droplet size is 0.01in to 0.04in (0.3mm to 1.0mm), and the best results are obtained when the droplets are of uniform size. At a diameter of 0.01in, the surface area is increased by a factor of 1,400/1 over solid water of the same volume.

Thus, little drops of water projected by a fog nozzle provide a heat-absorbing capacity far greater than that of the water projected by a smooth-bore nozzle. This capacity is certainly hundreds of times greater and in fact may be as much as a thousand times greater. This fact alone is sufficient reason to justify the statement that fog nozzles are far better than smooth-bore nozzles.

There is a second scientific fact that governs the rate of heat transfer. The rate of heat transfer depends upon the velocity of water near the burning surfaces and the fire plume. There must be sufficient velocity for water to be distributed evenly throughout the fire area. Even distribution is the second key principle for effective firefighting.

Using a smooth-bore nozzle, it is extremely difficult, if not impossible, to distribute water evenly throughout a confined space. The reach is much too great for average-size rooms. This results in splattering water off the walls or ceiling. With such a narrow stream, considerable time and movement are required to cover the entire space–all the while splattering the opposite wall. This is uneven distribution for sure.

The contrast with the use of fog nozzles is startling. The reach can be adjusted easily to extend just to the opposite wall. Shortening the reach widens the shape of the stream so that water can be distributed evenly within a few seconds. A single rotation can cover the entire space. This rotation should be continued until all the flames are blanked out and condensing steam appears. This is quite a contrast indeed.

Thus, on both key principles the fog nozzle is far superior to the smooth-bore nozzle. Is it any wonder that Layman stated, "Little if any progress can be made toward improving the tactical employment of water in firefighting operations until the fire service recognizes the gross inefficiency of the solid stream form of application."[8]

The conclusion from all of this is that smooth-bore nozzles are grossly inefficient in providing heat-absorbing capacity and even distribution. On the other hand, fog nozzles are highly efficient in meeting both principles.

Nozzle Reaction

The third consideration in comparing the two types of nozzles is nozzle reaction. Newton's Third Law of Motion states, "For every action there is an equal and opposite reaction." Newton's law applied to nozzles states that at equal pressures, a higher volume for one nozzle produces a greater reaction for that nozzle. Also at equal flows, a greater nozzle pressure produces a greater reaction. Since smooth-bore nozzle pressure is 50psi (3.47bar) while fog nozzle pressure is 100psi (6.9bar), should the nozzle reaction of a fog nozzle be twice that of a smooth-bore nozzle at equal flows?

The simple answer to this question is "no". For example, a 1.75in (44mm) tip flowing 118gpm (446Lpm) has a nozzle reaction of 44psi (3bar). A 100gpm (378Lpm) flow through a fog nozzle has a nozzle reaction of 51psi (3.52bar). The reaction force of 44psi compared with 51psi is close enough so there should be no significant difference in handling either nozzle. So it appears that flow is more important in determining nozzle reaction than nozzle pressure.

Let's examine the exact relationship between nozzle reaction, nozzle flow, and nozzle pressure. For example, let's look at a non-automatic nozzle rated at 100gpm (378Lpm) at 100psi (6.9bar). To increase the flow to 110gpm (416Lpm), a nozzle pressure of 120psi (8.3bar) is required. For a flow of 120gpm (454Lpm), the required nozzle pressure is 145psi (10bar). At 130gpm (492Lpm), the nozzle pressure is 170psi (11.7bar).

To calculate nozzle reaction for a given pressure and flow, the following formula is used:

$$NR = (0.0505)(Q)(\sqrt{NP})$$

where:

NR = nozzle reaction (pounds or kilograms)

Q = flow (gallons per minute or liters per minute)

NP = nozzle pressure (pounds per square inch or bars)

0.0505 = a constant

Note that the square root of the nozzle pressure is used. The square root of a number is much smaller than the number itself. This means that the flow is much more important in determining nozzle reaction, while the nozzle pressure is much less significant. So the difference in nozzle pressure between smooth-bore and fog nozzles is much less important when determining nozzle reaction. With this formula we can calculate the nozzle reaction for each of these flows:

100gpm = 50lb NR   378Lpm = 18.6kg

110gpm = 61lb NR   (+ 10% flow, + 22% NR)   416Lpm= 22.8kg

120gpm = 75lb NR   (+ 20% flow, +46% NR)   454Lpm = 28kg

130gpm = 86lb NR   (+30% flow, + 72% NR)   492Lpm = 32kg

With this non-automatic fog nozzle, nozzle reaction increases twice as fast as the flow.

Smooth-bore nozzles are subject to the same rules for nozzle reaction as non-automatic fog nozzles. Both types have a fixed tip size. The same nozzle reaction formula can be used with smooth-bore nozzles, but the formula is commonly rewritten as follows:

$$NR = (1.57)(d^2)(NP)$$

where:

1.57 = a constant

d = nozzle diameter (inches or millimeters)

A 1in (25.4mm) tip smooth-bore nozzle has the same nozzle reaction increases as the 100gpm non-automatic fog nozzle. For example, [9]

210gpm = 79lb NR   795Lpm = 29.5kg

230gpm = 94lb NR   (+ 10% flow, + 19% NR   870Lpm = 35kg

249gpm = 110lb NR   (+ 19% flow, + 39% NR)   942Lpm = 41kg

266gpm = 126lb NR   (+ 27% flow, + 59% NR)   1,006Lpm = 47kg

Smooth-bore nozzle reaction also increases at a rate about twice as fast as flow increase. This is a substantial increase in nozzle reaction for a marginal increase in flow rate. Remember why this occurs: nozzle pressure is multiplied by the increase in flow since the tip size cannot change. So the product of the increase in pressure times the increase in flow results in a greater percentage increase in nozzle reaction.

Automatic fog nozzles have the capability of keeping the increase in nozzle reaction to a minimum for any give flow by maintaining constant nozzle pressure. A non-automatic 250gpm (946Lpm) fog nozzle on a 2.5in (63.5mm) hose exhibits the same reaction characteristics as the 100gpm (378Lpm) fog nozzle if it is operated above its rated flow.[10]

250gpm = 126lb NR   (100psi NP)   946Lpm = 47kg

275gpm = 153lb NR   (+10% flow, +21% NR)   1,040Lpm = 57kg

300gpm = 182lb NR   (+ 20% flow, +44% NR)   1,135Lpm = 68kg

325gpm = 214lb NR   (+30% flow, +70% NR)   1,230Lpm = 80kg

Using the same flows, the nozzle reaction for an automatic nozzle is

250gpm = 126lb NR (100psi NP)   946Lpm = 47kg

275gpm = 139lb NR (+ 10% flow, + 10% NR)   1,040Lpm = 52kg

300gpm = 152lb NR (+ 20% flow, +21% NR)   1,135Lpm = 57kg

325gpm = 164lb NR (+30% flow, +30% NR)   1,230Lpm = 61kg[11]

Note that the percentage increase in nozzle reaction is equal to the percentage increase in flow (gallons per minute, liters per minute). A 10% increase in flow produces a 10% increase in nozzle reaction, and so on. Thus, the automatic fog nozzle is superior to other types of nozzles in providing a minimum increase in nozzle reaction for increased flows.

As the nozzle reaction increases with any nozzle, an equal or greater amount of counter-reaction must be produced to keep the nozzle stationary. In most firefighting operations the firefighting team supplies this counter-reaction. Just how much flow can a team handle? From past experience and experimentation, flow rates of 150gpm to 250gpm (568Lpm to 946Lpm) are workable volumes for automatics with a two-person team.

Should the nozzle reaction become excessive for a lone operator, the Task Force Tip automatic fog nozzle is the only nozzle that allows the operator to adjust the flow and the nozzle reaction without affecting the nozzle pressure or the stream quality. Operator flow control is an important safety factor to be considered whenever a nozzle is used.

Two Nozzle Myths

Is the following statement true or false: If a smooth-bore nozzle and a fog nozzle are both flowing the same amount of water, then the solid stream from the smooth-bore nozzle will hit harder than the straight stream from the fog nozzle.

A widespread myth in the fire service is that a solid stream from a smooth-bore nozzle penetrates better than a straight stream for a fog nozzle. Think about a bullet leaving a rifle. If two bullets leave different types of guns yet leave at the exact same velocity and with the exact same weight, they will go the same distance and hit with the same force. Once the bullet leaves the gun, it matters only how fast it is going and how much it weighs (kinetic energy).

In fact, a fog nozzle does a better job of focusing the stream at close range. The stream from a smooth-bore nozzle starts to spread the instant it leaves the nozzle. It must, because there is nothing to contain it. The fog nozzle, on the other hand, by the nature of its design, takes the water out at an angle and turns it so that at a distance of about 15ft (4.5m) the fog nozzle is putting more water through a smaller circle than a smooth-bore nozzle would. To be sure, the difference is small, but it is measurable.

The important thing to remember is that there is a common belief that smooth-bore streams are "solid" and hit harder. A fog nozzle straight stream is every bit as solid as a smooth-bore stream once it is measured past its focal point. Past the focal point it is simply a comparison of mass (water flow rate) and velocity (pressure) that determines impact force. In fact, a straight stream from a fog nozzle will have greater impact and penetration since it operates at 100psi (6.9bar) pressure rather than 50psi (3.4bar) for smooth-bore nozzles.

What about this statement? True or false: The straight stream from a fog nozzle is "hollow" as opposed to a solid stream from a smooth-bore nozzle.

The answer is false. It is easy to understand why this is a common belief. Anyone who has looked at a fog nozzle and understands how water comes out knows that at least for some distance the stream is hollow. However, it is hollow only for a very short distance–a distance that is much less than what would be used for firefighting. In fact, the hollow part is not filled with air but is a vacuum. This vacuum is very important for stream quality.

Water is accelerated out a distance; then it turns the corner and exits the nozzle parallel to the nozzle centerline. The vacuum in the center pulls in equally on all sides. In a very short distance the force of the vacuum actually causes the stream to pull together and become a compact stream. This is very easy to prove to yourself. Take two nozzles that are flowing the same amount of water–one fog nozzle and one smooth-bore nozzle. Using a pitot gauge, measure the pressure about 3ft from the exit from the nozzles. You will be able to easily pitot both streams across the stream. This means the stream from the fog nozzle is not hollow and is every bit as solid past the focal point as a solid stream from a smooth-bore nozzle.

This ends the comparison of fog nozzles and smooth-bore nozzles. Fog nozzles are much more efficient with a far greater capacity to absorb heat and to evenly distribute water throughout the fire area. In addition, automatic fog nozzles are capable of minimizing the increase in nozzle reaction resulting from increased flows above standard flows.

Now let's focus on fog nozzles and compare the various types and components of fog nozzles.

Fog Nozzle Teeth

All fog nozzles are capable of shaping the fog stream from a straight stream for reach and penetration to a wider fog pattern for greater heat absorption, firefighter protection, and other applications. However, there is more to shaping a fire stream than just turning the bumper. Most fog nozzles rely on some form of teeth to break water into droplets.

The earliest fog nozzles had square-faced metal teeth. Two problems existed: (1) the square-faced teeth left gaps or "fingers" in the fog pattern that allowed heat to pass through and (2) the metal teeth were susceptible to damage when dropped or used as a forcible entry tool. (See Figure 8–5)

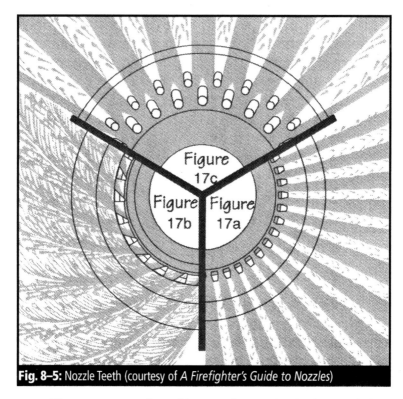

Fig. 8–5: Nozzle Teeth (courtesy of *A Firefighter's Guide to Nozzles*)

The next generation of fog nozzles used spinning teeth that appeared to eliminate the fingers in the fog pattern The spinning teeth were often made of plastic and were easily damaged or broken. High-speed photographs showed that the fingers were still there, even though they were not visible. The result was a wider, thinner fog pattern that spread the available water out beyond what was needed. Maximum width was just wide enough to cast a dense shadow of protection for the nozzle team. In addition spinning teeth produced extremely fine droplets that could be carried away rapidly. Finally, spinning teeth did not direct water to the center of the pattern. In fact, they prevented water from filling back into the middle of the fog pattern, so it remained hollow. This caused hot gases and flames to be sucked toward the face of the nozzle.

A later development used double rows of teeth. This was an attempt to fill in the gaps between the first row of teeth by creating another point of deflection. However, the second row formed "fingers" of its own and still left gaps in the pattern. An innovation used molded rubber teeth as an integral part of the bumper. The strong plastic teeth resisted damage by springing back to their original shape after impact. The thick rubber bumper helped protect these teeth. The rubber teeth were designed to produce a wide range of droplet sizes, from moderately coarse to extremely fine. This pattern had maximum heat absorption due to the fine droplets yet produced large droplets for maximum reach and penetration. The combination of these two effects provided a densely filled cone of water.

Shut-off Valves

Many fog nozzles in service today use a ball shut-off valve. However widespread their use, these nozzles have problems. Because they are designed to be operated in a fully open or fully closed position, partial opening produces a violent turbulence within the nozzle which destroys the straight stream and results in surging that disrupts the wider fog pattern. As nozzle pressure increases, the ball valve becomes more difficult to open because the ball is forced harder and harder against the valve seat.

Elkhart Brass advertises that its shut-off valve is hydraulically balanced, so presumably this solves the problems associated with ball valves. Task Force Tips uses a slide shut-off valve for its handline nozzles. This valve design controls the flow without creating turbulence. The pressure control unit compensates for the change of flow by moving the baffle to adjust to the proper tip size, maintaining correct nozzle pressure. Because of this action, the slide shut-off valve allows the nozzle to be operated at any handle position without producing turbulence. The stainless steel slide valve will not bind or tighten with age, wi'' not tighten under high pressure, and is always easy to open. A turbulence-free slide valve with automatic pressure regulation adds up to nozzle operator flow control.

Hydraulics; Conventional Nozzles

In order for a smooth bore or non-automatic fog nozzle with a fixed opening (conventional nozzle) to operate with the correct nozzle pressure and the proper flow, the correct pump discharge pressure must be supplied. Setting the correct pump pressure must include consideration of the available water supply, hose size, hose length, and pump capacity. If all of these things go right, a given flow of water passes through the nozzle at the desired nozzle pressure to produce an effective fire stream. This is a big "if".

If the proper fire stream is attained, the flow to that nozzle cannot be changed unless the discharge opening is changed (manually adjusted) for the new flow with a corresponding adjustment in pump discharge pressure. Since conventional nozzles cannot change tip size, one of two things must happen.

First, if a conventional nozzle is supplied less than the rated flow, the result is a weak, ineffective stream. This may be due to poor water supply, long hose lays, improper selection of tip size, or pump operator error. This under pressurized stream may waste water because the velocity needed to reach the fire is not produced. Little, if any, knockdown capability is achieved. Poor water supplies are often blamed for poor fire streams. More often, poor streams result from the inability to match the correct nozzle size to the available water supply.

Second, if more than the required flow is being delivered to a conventional nozzle, excessive nozzle pressure results. This excessive flow produces a proportionally higher nozzle pressure and therefore a corresponding increase in nozzle reaction or kickback. The higher nozzle reaction makes the hose line more difficult to handle and may jeopardize the safety of the nozzle team in an environment that is already unsafe.

Any attempt to control the over pressured line by the nozzle operator's cutting back at the nozzle results in a fire stream that is broken and erratic. A partially open ball valve creates tremendous turbulence that reduces the stream's effectiveness.

Hydraulics; Automatic Nozzles

The simplified hydraulics of the automatic nozzle can be easily remembered as the water triangle (Figure 8-6). Each side of the triangle represents one of the three limits to any pumper layout: water supply, pumper power, and maximum pressure. Working the pumper to whichever of the limits is reached first produces the maximum delivery for that layout.

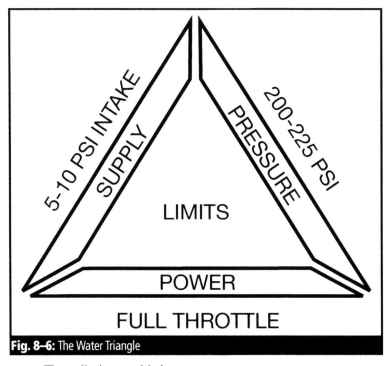

5-10 PSI INTAKE

SUPPLY

200-225 PSI

PRESSURE

LIMITS

POWER

FULL THROTTLE

Fig. 8–6: The Water Triangle

These limits would show as:

1. Water Supply–indicated by 5psi to 10psi (0.3bar to 0.68bar) on the inlet gauge or by the section hose going slightly soft. It can also show as the engine tending to run away.

2. Power–indicated by running out of throttle.

3. Pressure–indicated by the limiting pressure, usually 200psi to 225psi (13.8bar to 15.5bar) showing on the pumper discharge pressure gauge.

While these limits will yield the maximum for a particular layout, this is not to say that the layout shouldn't be improved. If you are working against the pressure limit, adding parallel or larger diameter lines can greatly increase flow. If water supply is the problem, improvement is necessary on the suction side of the pump. This can be accomplished with larger suction lines or additional lines into the pump or by receiving water from an additional source, such as relay pumping. The power limit is reached only at high volume–usually when supplying overcapacity to one or more streams. The load can be shared with a second pumper by shifting lines. The second pumper can pump in tandem off the same hydrant with the first pumper. Additional parallel or larger diameter lines can be used to reduce friction loss.

Although the same limits apply to a pumper when working with conventional nozzles, merely working to the system's limit does not produce desired results unless the tip size is exactly correct. If the regular tip size is too large, a poor under-pressurized stream is all that can be obtained. If the regular tip is too small, the stream will be over-pressurized, failing to deliver the volume available using the correct tip size.

With automatic nozzles, the pump operator can achieve maximum efficiency as fast as he can adjust the throttle. The automatic nozzle simultaneously adjusts the "tip size" to best deliver the available water. Scientific laws guarantee the best results faster and more accurately than with conventional nozzles.

So the second advantage of automatic fog nozzles is the ability to provide maximum effective fire streams at 100psi (6.9bar) nozzle pressure even though the pump pressure varies from that required for the standard nozzle flow. The automatic nozzle varies the flow to maintain the ideal constant pressure at the nozzle.

Hydraulics Formula

To demonstrate the flexibility of automatic nozzles, let's introduce the formula for calculating engine pressure:

$$PDP = NP + TPL$$

where TPL is total pressure loss, or appliance friction loss plus hose friction loss plus elevation pressure loss, PDP is pump discharge pressure, and NP is nozzle pressure.

In the following calculations, we assume that both appliance friction loss and elevation pressure loss equal zero. With these two factors at zero, TPL = FL (hose friction loss). Thus, the formula can be written as follows with NP = 100psi (6.9bar) for automatic nozzles:

$$PDP = 100 + FL$$

The formula for calculating FL is

$$FL = (C)(Q^2)(L)$$

where:

C = the coefficient for a given hose size

Q = flow in hundreds of gallons per minute

L = hose length in hundreds of feet

Our final version of the PDP formula then becomes

$$PDP = 100 + (C)(Q^2)(L)$$

As an example, what engine pressure is required to flow 300gpm (1,136Lpm) at 100psi (6.9bar) through 500ft (152.5m) of 2.5in (63.5mm) hose?

$$PDP = 100 + (2)(3^2)(5)$$

With $C = 2$ (coefficient table), $Q = 3$ (hundreds of gallons per minute), and $L = 5$ (hundreds of feet of hose),

$$PDP = 100 + 90$$

$$PDP = 190\text{psi} \qquad PDP = 13\text{bar}$$

Rules of thumb, charts, or tables may be used to find friction loss for a given flow or to find the flow for a given friction loss. (See Figure 8–7)

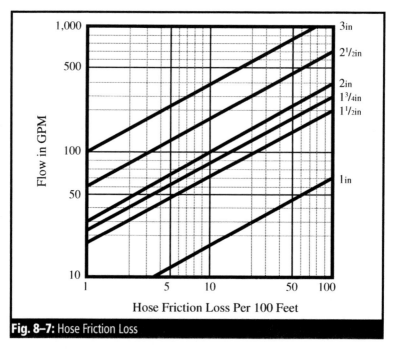

Fig. 8–7: Hose Friction Loss

What pressure do we pump to automatic nozzles? Automatic nozzles greatly simplify pump operation. Since automatic nozzles are designed to operate at 100psi (6.9bar), this becomes the starting point for any operation. Take an example: What pump pressure is required to flow 150gpm (568Lpm) through 200ft (61m) of preconnected 1.75in (44mm) hose? With C = 15.5, Q = 1.5, and L = 2, substituting into the FL formula gives

$$PDP = 100 + (15.5)(1.5^2)(2)$$

$$PDP = 100 + 70$$

$$PDP = 170\text{psi}$$

The advantage of TFT automatic nozzles is that any flow can be delivered by the pump operator and still be controlled by the nozzle operator. Variable flow, constant nozzle pressure, and nozzle operator flow control are three essential elements to successful fire streams and fire attack.

The following statements summarize the information presented thus far:

1. All fog nozzles provide much greater heat-absorbing capacity and much greater ability to distribute water evenly throughout the fire area. Thus, fog nozzles are highly efficient compared to the gross inefficiency of smooth-bore nozzles.

2. Automatic fog nozzles minimize the increase in nozzle reaction caused by increasing flows above the standard flow for that nozzle and hose size.

3. The TFT automatic nozzles provide a slide type shut-off valve that varies flow rates with no turbulence. The ball type shut-off valves cause extreme turbulence when they are partially opened.

Variable Flows

Two types of nozzles are capable of varying flow. The first type is the non-automatic nozzle with a manual volume control ring. Akron Brass offers the 1.5in (38mm) Turbojet nozzle with flows of 30gpm, 60gpm, 95gpm, and 125gpm (115Lpm, 230Lpm, 360Lpm, and 475Lpm). Akron Brass also offers a Wide-Range Turboject nozzle with a 1.5in (38mm) inlet with flows of 30gpm, 95gpm, 125gpm, 150gpm, and 200gpm (115Lpm, 360Lpm, 475Lpm, 550Lpm, and 750Lpm). Both nozzles are capable of reducing flows well below 100gpm (379Lpm) to make a balanced fire attack on room-size fires.

Elkhart Brass Company offers the Select-o-Flow nozzle with 1.5in (38mm) base with flows of 40gpm, 60gpm, 95gpm, and 125gpm (150Lpm, 230Lpm, 360Lpm, and 475Lpm). These nozzles match the Akron Brass nozzles in the flow selection below

100gpm (379Lpm). None of these nozzles is automatic. The Task Force Tips (TFT) automatic nozzles, like all automatic nozzles, maintain a constant pressure of 100psi (6.9bar) within the designed flow range. In addition the TFT automatics have a slide shut-off valve (instead of a ball valve) that does not produce any turbulence if the nozzle is partially opened. To help the nozzle operator, the TFT nozzles have six detent stops on the handle to provide lower flows than the maximum if the nozzle is fully opened.

This unique nozzle feature of the TFT automatic has two uses. First, the nozzle operator can use the detent stops to reduce flows to control nozzle reaction. No further action is required by the pump operator to do this. This action does not result in any deterioration of the fire stream. This is truly nozzle operator flow control with the important purpose of protecting the safety of the nozzle team. Second, the six detent stops can provide flows from 35gpm to 50gpm (132Lpm to 189Lpm) at detent stop one, on up to 150gpm to 200gpm (568Lpm to 757Lpm) for detent stop six.

Why is the ability of a fog nozzle to vary the flow so important? We fight fire with water. Water controls fire by absorbing the heat produced by the fire. Almost all of this heat-absorbing capacity occurs when liquid water is transformed to steam (gas) at 212°F (100°C). This means that we fight fires with steam. When steam is formed, it expands very rapidly. The expansion ratio is 1,700:1– a powerful force. It takes very little water to fill a room full of steam, in fact, no more than 5gal to 10gal (19L to 38L). So firefighters must handle this powerful blast of steam very carefully to produce an effective fog attack.

In essence, a safe and effective fog attack requires balancing the heat-releasing energy produced by the fire with the heat-absorbing energy produced by steam. Prior to the fog attack, the combustion process exists in a state of thermal balance. Immediately after an effective fog attack, this thermal balance is restored. This is called a balanced fire attack.

To make a balanced fire attack, you must apply the right amount of water to the fire–not too little and not too much. This means that all of the heat released by the fire is used to transform liquid water to steam and that all of the water is transformed to steam. This is what a balanced fire attack means. Such an attack controls a fire in the fastest possible time, with the least amount of water, and with minimum risk to the firefighters involved.

There are two very important implications that follow from the principle that you must use the right amount of water for an effective fog attack.

First, the right amount of water varies (linearly) as the size or volume of the fire. If fire fully involves a confined space, or room, then the volume of the fire equals the volume of the confined space. Suppose a second fire has twice the volume of a given fire. The right amount of water for the second fire will be twice that of the first (given) fire. Thus, fog nozzles must be able to vary the amount of water applied to fires. This can be done by varying the rate of flow as well as the time of application.

Second, the rate of flow must be below 100gpm (378Lpm) for room-size fires. No more than 5gal to 10gal (19L to 38L) is the right amount of water for such fires, and the rate of flow must be 30gpm to 60gpm (114Lpm to 227Lpm) to meet the tactical requirements for an effective fog attack.

The conclusion is that certain types of fog nozzles are unsuitable for a safe, effective fog attack. These are the fog nozzles with the ball shut-off valve and one rate of flow only–wide open. The flow rate of 100gpm (378.5Lpm) on up to 150gpm (571Lpm) is far too much for room–size fires. This is where fire departments throughout the US have run into problems in trying to make effective fog attacks. By using entirely too much water, they have created massive thermal imbalance problems. Without research to guide them, they do not realize the real cause of what is happening.

Only certain types of fog nozzles are suitable for a safe and effective fog attack. These nozzles are of three types. The first type of manual volume control nozzles are represented by the Akron Brass nozzle and the Elkhart Brass Turbojet and Select–o–Flow nozzles. The second type is the TA Fogfighter nozzle, popular in Europe and produced in Sweden, which has two flow rates with a ball shut–off valve that does not produce any turbulence for these two flows. (See Figure 8–8) The third type is the TFT automatic nozzle with a slide shut–off valve and six detent stops. These nozzles can be used to make a safe and effective fog attack.

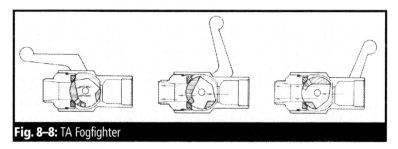

Fig. 8–8: TA Fogfighter

Blitz Attack Equipment

Throughout this book, emphasis has been placed upon the need to reduce flows for smaller room–size fires. What about larger fires requiring larger flows? The question of larger attack lines and higher flow rates has received a lot of publicity. Should we use them? Certainly, with the increase in hose diameter, there will be a marginal increase in the weight and size of an attack line. The larger attack line has two overwhelming advantages: firepower and time.

The 1.75in (44mm) or 2in (51mm) line, while retaining most of the handling benefits of the 1.5in (38mm) line, approximates or exceeds flows usually achieved with the 2.5in (63.5mm) line. Let's take an example. Suppose a certain fire requires 125gal (473L) of water to absorb all of the heat to control it. This is the right amount of water for a fully involved average–size house. With a booster line at 25gpm (94.6Lpm), 5min or more will be necessary. A standard 1.5in (38mm) line flowing 125gpm (473Lpm) requires 1min. A 1.75in (44mm) line flowing 250gpm (946Lpm) requires only 30s.

Nozzles with nozzle operator flow control in combination with larger size preconnects add a new dimension to the term "fire attack". Big line flows with fast, small hand lines are now available. Well–trained firefighters and teamwork are still required, but available personnel can now hit faster with more attack capability. With 1.75in (44mm) line at 200psi (13.8bar) pump pressure, flows are available up to the following:

230gpm (870Lpm) on 150ft (46m) preconnects

200gpm (757Lpm) on 200ft (61m) preconnects

175gpm (662Lpm) on 250ft (76m) preconnects

These flows are practical maximums for this size of line. With nozzle operator flow control, these flows can be reduced at the nozzle if the need arises.

At first, one may cringe at such high flow rates. "But we've got only 500gal (1892L) in our tank." Just 2min on one line, and 1min with a pair. With training and experience, you do not have to fear losing your water supply. After a quick preconnect stretch to the center of the fire, a blitz attack flow is delivered. A 10s to 15s, blast produces tremendous effect with steam penetrating the same areas as the fire. A room of intense fire can be quenched with far fewer than 50gal (189L) of water.

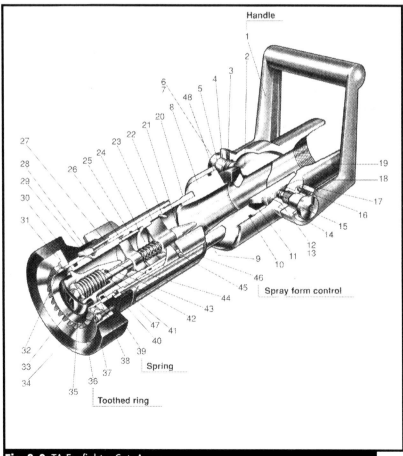

Fig. 8–9: TA Fogfighter Cut–Away

A booster line wouldn't have slowed the fire. A conventional 1.5in (38mm) line might have taken several minutes and several hundreds of gallons of water for control. Even 500gal (1892L) of tank water spread out over the first critical minutes produces unbelievable results. By that time a supply line from a hydrant or a tanker should be connected. If this blitz attack hasn't controlled the fire or at least bought time to supplement the supply, there is no way that an effective fire attack could have been made with smaller lines. Today's modern fog nozzles have the capability of handling the larger flows for a blitz attack as well as the smaller flows needed for smaller fires.

Finally, this chapter ends with the cut–away drawings of two fog nozzles. You can see the mechanics of each nozzle, that is, how they are designed to produce the little drops of water, and at the same time vary the flow so that the right amount of water can be used. Truly these nozzles are the right tools for a safe and effective fog attack.

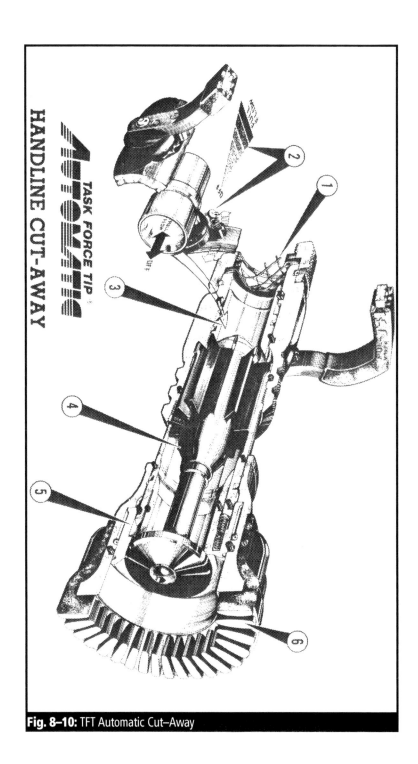

Fig. 8–10: TFT Automatic Cut–Away

Notes

[1]David P. Fornell, *Fire Stream Management Handbook* (Saddle Brook, New Jersey: Fire Engineering, 1991), p. 106.

[2]Harry R. Hickey, *Hydraulics for Fire Protection* (Quincy, Mass.: NFPA, 1980), p. 221.

[3]*A Firefighter's Guide to Nozzles* (Valparaiso, Ind.: Task Force Tips, 1995), p.6.

[4]Ibid, p. 10.

[5]Ibid, p. 11.

[6]Ibid, p. 10.

[7] Layman, *Attacking and Extinguishing Interior Fires,* Quincy, MA, National Fire Protection Association, 1955), p. 20

[8] Ibid, p. 26.

[9]A Firefighter's Guide, op. cit., p. 29.

[10]A Firefighter's Guide, op. cit., p. 30.

[11]A Firefighter's Guide, op. cit., p. 31.

[12]A Firefighter's Guide, op. cit., p. 23.

[13]A Firefighter's Guide, op. cit., p. 41.

[14]A Firefighter's Guide, op. cit., p. 19.

[15]Tour & Andersson Hydronics AB, *Fogfighter* Ljung, Sweden: Tour & Andersson Hydronics (1994), p. 3.

[16]Ibid, p. 6.

[17]*A Firefighter's Guide,* op. cit., p. 67.

The General Rate–of–Flow and Grimwood Formulas

Now the authors want to present the most significant result of the research done in the United States and in Europe. This result is the proof that two Rate–of–Flow formulas are actually the same.

Keith Royer and Bill Nelson created the Iowa gallonage formula at the Fire Service Institute at Iowa State University in 1954. The formula is:

$$gal = \frac{Vol}{200}$$

In this equation "gal" equals the number of gallons of water (the right amount of water) needed to fill a confined space full of steam, "Vol" equals the volume of the confined space in cubic feet.

To convert the Iowa gallonage formula to a metric formula, we must start with the liter. The liter, like a gallon, is a measure of volume with one gallon equal to 3.785L. One liter equals on cubic decimeter, or one one–thousandth of a cubic meter. Since the volume of structures is usually expressed in cubic meters, we must transform liters to cubic meters. This is easy to do with the metric system.

The expansion ratio of liquid water to steam is 1,700/1 no matter what unit of measure is used. Therefore, one liter of water expands to 1,700L of steam. The liter formula is:

$$L = \frac{V}{1,700}$$

In this equation both L and V are in liters. To change the number V to cubic meters, it is necessary to divide the numerator and denominator of the fraction by 1,000.

$$L = \frac{V/1,000}{1,700/1,000}$$

V/1,000 may be rewritten as "Vol" in cubic meters. The denominator becomes 1.7.

$$L = \frac{Vol}{1.7}$$

In this equation "L" equals the number of liters, and "Vol" equals the volume in cubic meters. Since the Iowa formula assumes that 90% of liquid water is transformed to steam, we must do the same for the metric formula. Thus the denominator becomes 1.5, that is, 90% of 1.7. The equation is the general Rate–of–Flow formula in metric units.

Paul Grimwood's article[1] discusses his formula for minimum fire ground flow–rate requirements:

$$A \times 2 = Lpm$$

To transform Grimwood's formula into the General Rate–of–Flow formula, we add time to the formula and multiply by the liters per minute. There is a basic mathematical reason for this. The equation states, in essence, that a certain number of liters (A x 2) equals the same number of liters per minute. However, liters per minute is a rate of flow, and this equation is valid only if time equals 1min ((Lpm) x 1 = Lpm). This is the only time that will produce the same number of liters per minute on both sides of this equation. We don't want to be restricted to an attack time of 1min; we want an equation valid for any length of time. So making this change in Grimwood's formula gives

$$A \times 2 = \text{Lpm} \times t$$

This generalizes Grimwood's formula and makes it valid for any length of time.

To transform Grimwood's formula to volume, the area (A) must be multiplied by the ceiling height. Let's do this for ceiling heights of 8ft and 10ft (2.5m and 3m). Of course, if we multiply the numerator by 2.5 or 3, we must do the same for the denominator:

$$\frac{3 \times A \times 2}{3} = \text{Lpm} \times t \qquad \frac{2.5 \times A \times 2}{2.5} = \text{Lpm} \times t$$

Since 3 x A (or 2.5 x A) equals volume, we can change notation to get the following equations:

$$\frac{\text{Vol} \times 2}{3} = \text{Lpm} \times t \qquad \frac{\text{Vol} \times 2}{2.5} = \text{Lpm} \times t$$

Our final change is to simplify the fraction by eliminating the number in the numerator of each fraction. This is done by multiplying the numerator and denominator by 0.5:

$$\frac{\text{Vol}}{1.5} = \text{Lpm} \times t \qquad \frac{\text{Vol}}{1.25} = \text{Lpm} \times t$$

Note that the 10ft (3m) ceiling height gives an equation identical to the General Rate–of–Flow formula. The 8ft ceiling height (2.5m) is almost within 90% of the general formula.

What is the significance of this finding? First, these two formulas were created 36 years apart and in different countries. Second, both formulas are the result of careful research based upon scientific facts and principles. This convergence adds further proof to the validity of the General Rate–of–Flow formula. All of the research that has been done on fog nozzles has converged upon identically the same set of facts and principles. It is safe to say that the General Rate–of–Flow formula is the only valid formula the fire service will ever have to work with for confined fires.

Chief Lloyd Layman's Research

- Created the indirect method of fire attack using fog nozzles, the first alternative to the solid–stream attack from a smooth–bore nozzle.

- Created a theory of atmospheric displacement to explain how the indirect method worked.

- Successfully adapted the indirect method to fighting Class A structure fires.

- Demonstrated the indirect effect of a fog attack, whereby fire is controlled in adjacent areas remote from where the fog attack is made.

- Initiated the change from solid–stream nozzles to fog nozzles in this country in his "Little Drops of Water" speech in Memphis (1950).

- Presented an exposition of the theory and practice of fighting fires with water fog in his book, *Attacking and Extinguishing Interior Fires* (1955).

- Completely analyzed the eight tactical operations used in fighting fires in Tactics (1953).

Iowa State University Research
Keith Royer and Floyd W. (Bill) Nelson

- Created the gallonage formula, the source of the Iowa Rate–of–Flow formula. The gallonage formula answers the question, "How much water is needed for control of a confined fire?"

- Stated the fundamental principle of fire control that depends upon using the right amount of water.

- Discovered that using too much water causes thermal imbalance problems that disrupt an effective fog attack.

- Confirmed that using too little water has little effect upon a fire.

- Created the most complete and accurate analysis of uncontrolled fire behavior in a structure.

- Created the the combination attack with an efficiency of better than 90% conversion of water to steam.

- Stated the fundamental tactical principle that chooses the best method of attack for a given purpose and a given type of fire.

- Participated in creating the constant flow fog nozzle and the elevated master stream nozzles and aerial equipment to combat large fires.

- Derived sound conclusions from the scientific method of systems analysis that established a solid foundation for the safe and effective use of fog nozzles.

US Naval Research Laboratory

- Quantified the conditions required to produce the maximum compartment fire temperatures with natural ventilation.

- Defined two thermal conditions necessary and sufficient for flashover to occur: heat flux to the floor and upper layer temperature.

- Established the optimum application rate for boundary cooling together with the best ways to distribute water for horizontal and vertical cooling.

- Demonstrated the safety and effectiveness of 3–D gas cooling pulse tactics.

- Proved the superiority of 3–D gas cooling pulse tactics versus the straight–stream attack in providing a safe entry to a compartment when an initial direct attack is impossible to make.

- Demonstrated that a direct attack with a straight stream from a fog nozzle and a solid stream from a smooth–bore nozzle are equally effective.

- Established the importance of maintaining thermal balance when using fog nozzles.

- Demonstrated the weaknesses of protective clothing— especially gloves—that limit how long firefighters can remain in a hostile environment.

European Research

- Created 3–D gas cooling pulse tactics to enable firefighters to enter a burning compartment safely.

- Through the pulse tactic, eliminated the twin dangers of flashover and backdraft.

- Proved that the pulse tactic actually contracted the fire gases and steam in a compartment, preserving thermal balance.

- Conducted extensive research on all aspects of fog attack, including a comparison of various tactics, water droplet size, and fog patterns.

- Through research, confirmed the superiority of 3–D gas cooling tactics.

- Established the superiority of the automatic nozzle with slide shut–off valve.

- Created a flow rate formula that is identical to the General Rate–of–Flow formula.

Research Summary

All of the research on fog nozzles converges on the following set of facts and principles:

- The proper use of fog nozzles requires a balanced fire attack, that is, balancing the heat–releasing power of a fire with the heat–absorbing power of water (steam).

- Doing this requires using the right amount of water applied using the near–ideal rate of flow.

- A fog nozzle must be capable of varying the flow; for room–size fires the ideal rate of flow ranges from 30gpm to 60gpm (113Lpm to 227Lpm).

- Using the right amount of water results in a net contraction of fire gases and steam and the restoration of thermal balance.

- Thermal imbalance problems (*i.e.*, pushing a fire) are caused by using too much water.

- A straight stream from a fog nozzle, a narrow fog pattern (30°), and a wide angle pattern (60°) are useful in various fog attacks.

- No one method of fog attack is useful for all purposes and all types of fires.

- A fog nozzle is far superior to a smooth–bore nozzle in its ability to vary the flow and distribute water evenly throughout the fire area.

- All of this research leads to the final conclusion that a fog nozzle can be used safely and effectively provided it is used properly.

Summary

It is easy to see now why this book was written. Fifty years of research, completed in more than ten countries, all leads to the same set of conclusions:

- Fog nozzles can be used safely and effectively. In fact, these are the only nozzles that can handle certain types of fires.

- Proper use of fog nozzles can preserve thermal balance and provide safe entry for firefighters into the fire area.

- The formulas created in the English and metric systems are identical. Thus, there is only one valid formula to determine the needed fire flow for a fog attack.

- The direct attack, the combination attack, and the 3–D gas cooling pulse tactic are all useful for fire attack. No one method of attack can handle every fire situation.

Notes

[1]Paul Grimwood, "Compartmental & Structural Firefighting: Water Flow Requirements – International Research" (www.firetactics.com, 1999).

Bibliography

Babrauskas, Vyetnis "Burning Rates", SFPE Handbook of Fire Protection Engineers, Quincy MA: National Fire Protection Association, 1995.

Barnett, C.R., Macbar Fire Design Code, Auckland, New Zealand.

Bengtsson, Lars–Goran, Report 12019, Lund University 1999, Sweden.

Campbell, John A "Confinement of Fires in Buildings" National Fire Protection Association Handbook 17th Edition, Quincy MA: NFPA 1991.

Carlson, Gene "Lloyd Layman's Theory: Its Time has Come and Gone", *Fire Engineering*, Vol. 135 No. 2 Feb. 1983.

Chitty, Richard, Report 5–94, UK Research and Development Group, (FRDG) United Kingdom (Great Britain).

Clark, David "Let's Get Something Straight", Illinois Fire Service Institute, Champaign IL, Summer 1990.

Clark, David "Straight Talk about Nozzles and Fire Attack", Illinois Fire Service Institute, Champaign IL, unpublished newsletter.

Clark, Frederick B. "Fire Hazards of Materials – an Overview", National Fire Protection Association Handbook 17th Edition. Quincy MA: NFPA 1991.

Clark, William F. Firefighting Principles and Practices. Saddle Brook N.J.:" Fire Engineering 1991.

Comeau, Ed "Roof Collapse Kills Three", *National Fire Protection Association Journal*, Vol. 93 No. 4 July–August 1999.

Darwin, R.L. et al Post–Flashover Fires in Shipboard Compartments Aboard Ex–USS Shadwell, Phase VI, Boundary and Compartment Cooling. Mobile AL: Naval Research Laboratory 1994.

Davis, Larry *Rural Firefighting Operations*, Vol. 2. Ashland MA: International Society of Fire Service Instructors, 1986.

Farley, John P. et al Phase I – Full Scale Offensive Fog Attack Tests. Mobile AL: Naval Research Laboratory 1997.

Farley, John P. "Fog Attack for Ship Fires", *Fire Engineering*, Vol. 147 No. 7 July 1996.

Fire, Frank Combustibility of Plastics. New York: Van Nostrand Reinhold 1991.

Fitzgerald Robert W., "Structural Integrity during Fires". National Fire Protection Association Handbook 17th Edition. Quincy MA: NFPA 1991.

Fleischman, Charles, NIST–GCR–94–846, National Institute of Standards and Technology, 1994, University of California, Berkley, CA.

Fornell, David P. Fire Stream Management Handbook. Saddle Brood N.J.: Fire Engineering 1991.

Grant & Drysdale D.. Report 1/97 UK Fire Research and Development Groups (FRDG), United Kingdom (Great Britain).

Grimwood, Paul "New Wave 3–D Water Fog Tactics", *Fire Engineering* Vol. 153 No. 10 Oct 2000 p. 99.

Grimwood, Paul Fog Attack. Redhill UK: FMJ International Publications LTD, 1992.

Hickey, Harry R. Hydraulics for Fire Protection. Quincy MA: National Fire Protection Association, 1980.

Huggett, Clayton, "Estimation of Rate of Heat Release by Oxygen Consumption Measurements" *Fire and Materials*, Vol. 4 No. 2, Feb 1980, p. 54.

Kleene, Barnard J. and Sanders, Russell J. Structural Fire Fighting. Quincy MA: National Fire Protection Association, 2000.

Knapp, Jerry and Delisio, Christian, "Energy–Efficient Windows Firefighters Friend or Foe", *Firehouse*, Vol. 22 No. 7 July 1977, p. 74 ff.

Knopt, Richard A "Fog Streams Compared with Straight Streams", *Fire Chief*, Vol. 23 No. 7 July 1979, p. 36.

Lauren, Anders, personal communication, 1990.

Layman, Lloyd Attacking and Extinguishing Interior Fires. Quincy MA: National Fire Protection Association, 1955.

Leonard, J.T. et al, Post–Flashover Fires in Simulated Shipboard Compartments, Phase I Small Scale Studies, Chesapeake Bay Detachment Fire Test Facility, Naval Research Laboratory, 1991.

Lie, T.T. "Fire Temperature Time Relations", SFPE Handbook of Fire Protection Engineers. Quincy MA National Fire Protections Association, 1995.

Nelson, Floyd W (Bill) Qualitative Fire Behavior. Ashland MA, International Society of Fire Service Instructors, 1991.

Richman, Harold "Improving Interior Fire Attack", *Fire Command* Vol. 53 No. 7 July 1986, p. 18.

Royer, Keith, "Water for Fire Fighting", Iowa State University Engineering Extension Service. Ames IA: Bulletin 18, undated.

Royer, Keith "Test Fire for Exploratory Committee on Application of Water", Iowa State University Engineering Extension Service, 1959.

Royer, Keith "Report on Story City Fire Test" Iowa State University Engineering Extension Service, Bulletin, 1959.

Rosander, Mats and Gielson, Kristen, *Fire Magazine* UK, October 1984, p. 43 – 46.

Sardqvist, Stefan, Report 7003, Lund University, 1998, Sweden.

Scheffey, J.P., Siegmann, C.W., Toomey, T. A., Williams, F.W., Farley, J.P, Phase II – Full–Scale Offensive Fog Attack Tests, NRL/MR/6180–97–7944, U.S. Naval Research Laboratory, 1997.

Sutherland, B.J., Report 99/15, University of Canterbury, 1999 New Zealand.

Task Force Tips, A Firefighter's Guide to Nozzles, Valparaiso IN: Task Force Tips , 1995.

Tour & Andersson Hydronics AB, Fogfighter. Sweden Ljing: Tour & Andersson Hydronics, 1994.

Tuomisaari, Maarit, *Suppression of Compartment Fires with a Small Amount of Water* VTT (National Research Laboratory) Finland.

Wiseman, John D. The Iowa State Story, Stillwater OK: Fire Protection Publications, 1998.

Index

A

3-D pulse attack, 146

3-D water fog attack, xiii–xiv,
9–11, 71, 74–83
pulsating flow, xiii–xiv, 78
compartment fire, 76
gas cooling, 76–80, 83
atmosphere displacement,
76–77
fire suppression, 76–77
defensive/offensive tactics,
77, 81–83
ignition, 77, 81
simulation, 77, 80
contraction of gases, 78–80
pushing the fire, 79, 82
fog pattern, 79–80
heat transfer, 80
benefits, 81
oxygen level, 81
free radicals, 81
myths/misconceptions,
81–83
spray pattern, 82
personnel safety, 82
thermal balance/imbalance,
82
flow rate, 82–83
ventilation, 82

water-hammer effects,
82–83
amount of water needed, 83

Access to fire, xiii, 67,
120–122

Air draft, 4, 76

Air-intake opening, 2–3, 5,
14, 76
draft, 4, 76
size, 14

Akron Brass Turbojet nozzle,
142

Amount of water calculation,
133–138

Amount of water needed,
26–28, 33–36, 39, 68, 83,
118–120, 129, 132–138
fundamental principle,
26–28
counterproductive, 34–35
right amount, 39
calculation, 133–138

Application method, 39–41,
124–125, 144–150
fog pattern, 40
nozzle rotation, 40–41
attack methods, 124–125,
144–150